AMPHIBIAN EMBRYOLOGY

RECENT ADVANCES IN EMBRYOLOGY SERIES-2

Amphibian Embryology

A.P. DIWAN
&
N.K. DHAKAD

ANMOL PUBLICATIONS PVT LTD
New Delhi-110 002

ANMOL PUBLICATIONS PVT LTD
4374/4B, Ansari Road, Daryaganj
New Delhi-110 002

Amphibian Embryology

First Edition 1995
ISBN 81-7488-101-8

PRINTED IN INDIA

Published by J.L. Kumar for Anmol Publications Pvt. Ltd., New Delhi and Printed at Efficient Offset Printers, Delhi.

Preface

Embryology is a branch of biology which has a most immediate bearing on the problem of life. It has become an immense field whose confines are difficult to delineate. The focus has been and remain on the embryos, on the gradual emergence of form and structure from what appears to be a very modest beginning. The essence of embryonic development is transition from one stage to another. Embryo is a fleeting stage, a continuum along the axis of time.

The present volume *Amphibian Embryology* deals with the stages of metamorphosis in amphibians. Metamorphosis is a developmental adaptation permitting the change from water to land environment.

The authors have freely consulted other standard books and research papers while preparing the manuscript, so the authors claim no originality of the work.

Constructive criticisms and suggestions for improvement of the book will thankfully acknowledged.

Authors

Contents

1

Reproduction in Frog

THE MALE REPRODUCTIVE ORGANS

Sexual Characters

In frog sexual dimorplus is not well distinct. However during breeding season some characters appear. The female is biggcr the male. The identifying features of male which distinguish it from the female are a *darkened thumb pad* which changes thickness and colour intensity as the breeding season approaches; a distinct low, guttural *croaking sound* with the accompanying swelling by air of the lateral vocal sacs located between the tympanum and the forearm; a more slender and *streamlined body* than that of the female; and the *absence of coelomic cilia* except in the peritoneal funnels on the ventral face of the kidneys. Males of many species carry additional features such as brilliant colours on the ventral aspects of the legs (*R. sylvatica*), black chin (*B. fowleri*), or the size of colour of the tympanic membrane.

The Testes

There is a pair of testes present just above the dorsally placed kidneys and remain suspended by a double fold of peritoneum known as the *mesorchium*. This mesentery surrounds each testis and is continuous with the peritoneal epithelium which covers the ventral face of each kidney and lines the entire body cavity.

The testes are whitish and ovoid bodies lying ventral to and near the anterior end of each kidney. The *vasa efferentia*, ducts from the testes, pass between the folds of the mesorchium and into the mesial margin of the adjacent kidney. The number of vasa efferentia leaving the tesits is about a dozen. During the breeding season, or after slight compression of the testis of the hibernating frog these ducts become the more apparent due to the presence in them of whitish masses of spermatozoa in suspension. The ducts are very small in diameter, tough walled, and interbranching. They are lined with closely packed cuboidal cells. All vasa efferentia of one testis unit together inside the kidney of the related side and form a cerenal duct the *Bidders Canal* that transports sperms through the kidney. The Bidder canal is a longitudinal tube situated along the median edge of the kidney communicated with the writer by transverse collecting tubes. The presence of spermatozoa in the kidney also can be achieved artificially be injecting the male frog with the anterior pituitary sex-stimulating hormone.

The spermatozoa are produced in inside *seminiferous tubules*. These are closely packed, oval-shaped sacs, which are separated from each other by thin partitions (septula) of supporting (connective) tissue known as *interstitial tissue*. This tissue presumably has some endocrine function. The thickness of this tissue is much reduced immediately after breeding or pituitary stimulation. The interstitial tissue is continuous with the covering of the testes known as the tunica *albuginea*.

Spermatogenesis

When the photoperiodism is appropriate during breeding season the given cells undergo spermatogenesis but shortly after the breeding season the *spermatogonium*, which has ceased all mitotic activity, enters upon a period of rest but not inactivity. During this period the nucleus passes through a sequence of complex changes which represent an extended prophase. This is in anticipation of the two maturation divisions that finally produce the haploid spermatid which metamorphoses into a spermatozoon.

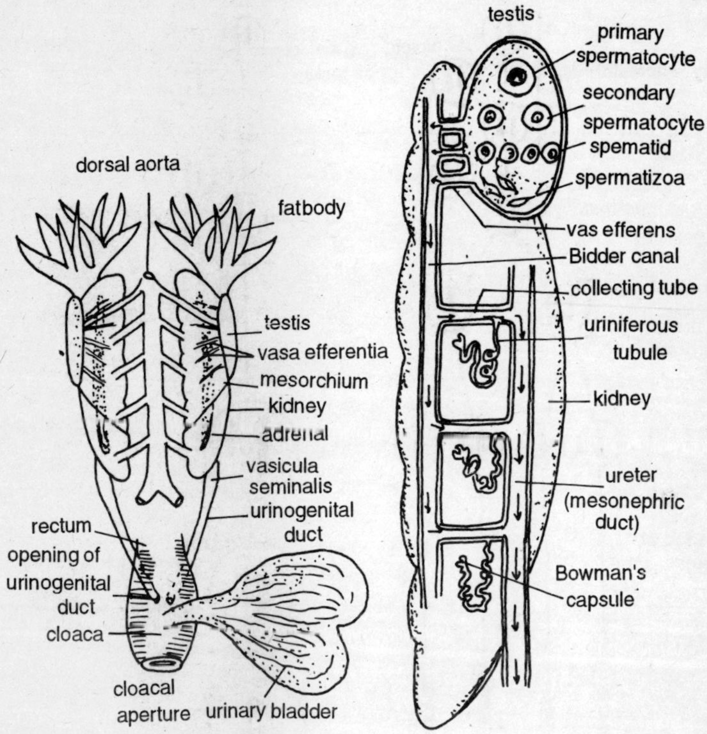

Fig. 1.1. Frog. Urinogenital organ of male

The Spermatogenia show several cytolosical charges. The nucleus of the spermatogonium contains chromatin which appears as relatively coarse lumps distributed widely over an achromatic reticulum. Both the cytoplasm and the nucleus grow and the chromatic granules become finely divided and arranged into contiguous rows, bound by an achromatic thread, together known as chromosomes. This is the *leptotene stage* of spermatogenesis. Shortly the chromosomes become arranged in pairs which converge toward that side of the nucleus where the centrosome is found. The opposite ends of the paired chromosomes merge into the general reticulum. This is the *synaptene stage* or *boughal stage*. The chromatin granules become

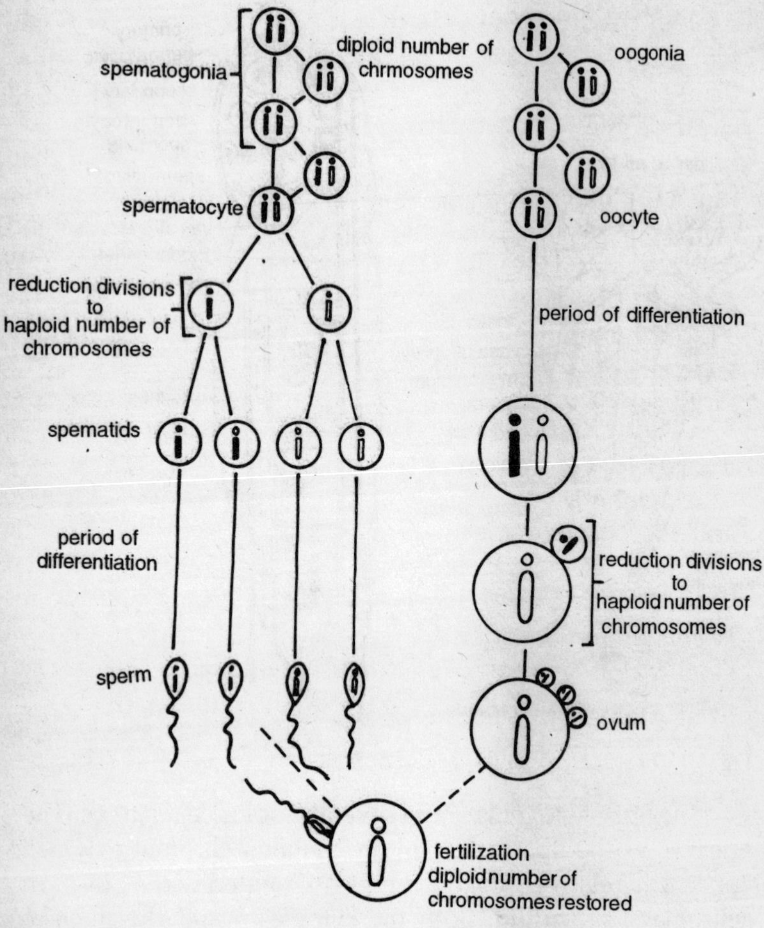

Fig. 1.2. Diagramatic representation of gametogensis.

telescoped together on the filaments so that the aggregated granules, known as *chromosomes*, appear much shorter and thicker. Pairs of chromosomes become intertwined and the loose terminal ends become coiled and tangled together. This is the *contraction* or *synizesis stage*. Then the members of the various pairs become laterally fused. While there is no actual reduction in total chromatin, there is a temporary and an apparent reduction in the total number of chromosomes to the haploid condition, because of this fusion. There is no actual reduction in the total amount of chromatin material, nor is there any permanent reduction at this stage in the number of chromosomes. Their identity is lost only temporarily. This is known as the *pachytene stage*. The members of each pair then separate again. It must be remembered, however, that (a) the separation need not be along the original line of fusion and that (b) an exchange of homologous sections of the chromosomes may occur without any cytological evidence. In any case, the diploid number of chromosomes reappears and this is then known as the *diplotene stage*.

During these changes in the chromatin material of the nucleus, the volume of the nucleus and the cytoplasm are considerably increased, the nuclear membrane breaks down, and the chromosomes assume bizarre shapes and various size. They may be paired, curved, or straight; "V" and "C" and reversed "L" shapes, figures 8's, and grouped as tetrads. This is known as the *diakinesis stage*. The chromosomes are then lined up on a spindle in anticipation of the first of the two maturation divisions.

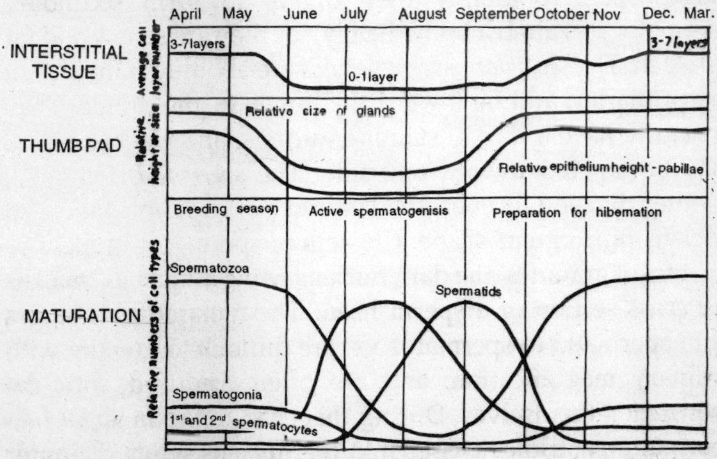

Fig. 1.3. Normal cyclic changes in the primary and secondary sexual characters of the frog, *Rana pipiens*.

Frog is a seasonal breeder and the spermatogenesis in is completed within the testes. The walls of the seminiferous tubules produce *spermatogonia* which go through mitotic divisions and then the series of nuclear charges without mitosis. This results in the appearance, toward the lumen of each tubule, of clusters of mature *spermatozoa*. The spermatogonia are found close to the basement membrane of the seminiferous tubule. These then await their turn to undergo the maturation changes necessary for the production of spermatozoa which will be ready for the breeding season. The elongated and filamentous tails of the clustered mature *spermatozoa* project into the lumen of each seminiferous tubule.

At the peak of spermatogenetic activity, all the stages of maturation from the spermatogonium to the spermatozoon are present. The spermatogonia are always located around the periphery of the seminiferous tubule and are small, closely packed cells, each with a granular, oval nucleus. In between the spermatogonia may be found occasional very large cells, the *primary spermatocytes*. These tend to be irregularly spherical, possessing large and vesicular nuclei. The cells are so large that they may be seen under low power magnification of the microscope. Apparently they divide to form secondary spermatocytes almost immediately, for they are so few and far between. The *secondary spermatocytes* are about half the size of the primaries, and lie toward the lumen of the tubule. They generally have a darkly staining nucleus, and the cytoplasm may be tapered toward one side. The *spermatid*, following another division, is even smaller and possesses a condensed nucleus of irregular shape. Clusters of spermatids appear as clusters of granules, the dark nucleus being almost as small as the cross section of a sperm head. The metamorphic stages from spermatid to spermatozoon are difficult to identify with ordinary magnification, and are often confused with the spermatids themselves. During this change the inner of two spermatid centrioles passes into the nucleus while the outer one gives rise to the tail-like flagellum.

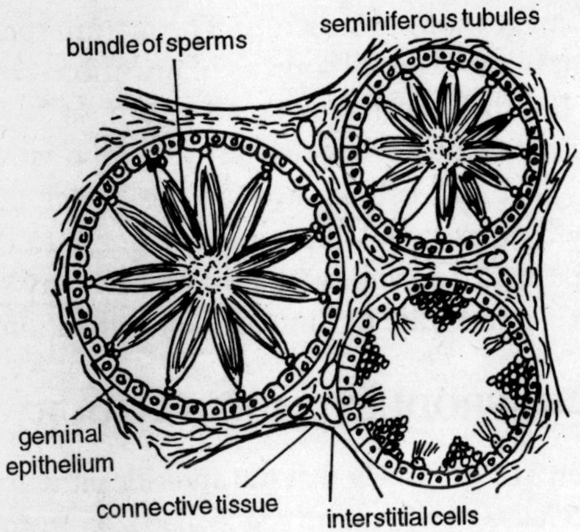

Fig. 1.4. Frog. T.S. of Testis.

The mature *spermatozoon* averages about 0.03 mm. in length. It has an elongated, solid-staining head (nucleus) with an anterior acrosome, pointing outwardly toward the periphery of the seminiferous tubule. The short middle piece generally is not visible but the tail appears as a gray filamentous extension into the lumen, about four or more times the length of the sperm head.

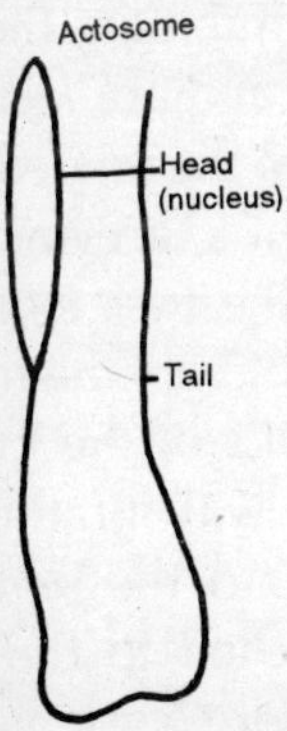

Fig. 1.5. Frog spermatozoon.

In any cross section of this testis, bundles of sperm heads or tails may be cut at right angles or tangentially, giving misleading suggestions of structure. The mature spermatozoon is dependent upon external sources of nutrition so that it joins from 25 to 40 other spermatozoa, all of whose heads may be seen converging into the cytoplasm of a relatively large, columnar-type basal cell known as the *Sertoli cell.* This is functionally a nurse cell, supplying nutriment to the clusters of mature spermatozoa until such time as they may be liberated through the genital tract to function in fertilization.

REPRODUCTIVE BEHAVIOUR

It has been proved definitely that the anterior pituitary hormone causes the release of the mature spermatozoa from the testis. But this hormone also releases other maturation stages. It is therefore probable that there are smooth muscle fibers, either among the interstitial cells or in the tunica albuginea of the testes, which fibers contract to force the spermatozoa from the seminiferous tubules. It would be as difficult to physiologically demonstrate the presence of these fibers in the testis as it is simple to demonstrate them in the contracting cyst wall of the ovary.

Responding to sex stimulation, the spermatozoa become free from their Sertoli cells and are formed from the lumen of the seminiferous tubule into the related *collecting tubule.* These collecting tubules are small and are linked with closely packed cuboidal cells. They join the *vasa efferentia* which leave the testis to pass between the folds of the mesorchium and thence into the *Malpighian corpuscles* of the kidney. From this point the spermatozoa pass by way of the excretory ducts, the uriniferous tubules, and into the *mesonepheric duct* (ureter) which may be found attached to the lateral margin of the kidney. Within the excretory system the spermatozoa are immotile, due to the slightly acid environment. They are carried passively down the ureter to the slight dilation near the cloaca, known as the *seminal vesicle.* Within the vesicle the spermatozoa are stored briefly in clusters until amplexus and oviposition occur. At oviposition the male ejaculates the spermatozoa into the neutral or slightly alkaline water where they are activated and then are able to fertilize the eggs as they emerge from the cloaca of the female.

During the normal breeding season amplexus is achieved as the females reach the ponds where the males are emitting their sex cells. During amplexus there are definite muscular ejaculatory movements on the part of the male frog, coinciding with oviposition on the part of the female. Amplexus may be maintained by the male for many days, even with dead females. As soon as the eggs are laid and the male has shed his sperm, he goes through a brief weaving motion of the body and then releases his grip to swim away. The frogs completely neglect the newly laid eggs.

ACCESSORY REPRODUCTIVE ORGANS

In the male frog the ureter is not directly connected with the *bladder*, as it is in higher vertebrates. It is possible that the bladder in the Anura may be an accessory respiratory and hydrating organ, particularly in the toads, where water may be stored during migrations onto land.

The male frog also has a duct, homologous to the oviduct of the female, known as the "rudimentary oviduct" or *Müllerian duct*. This duct normally has no lumen, and is very much reduced in size so that it may be difficult to locate. There is experimental evidence that this duct may be truly a vestigial oviduct since it respondents to ovarian or female sex hormones by enlarging and acquiring a lumen. At the anterior end of the testes of some Anura (e.g., toads) there may be found an undeveloped ovary known as *Bidder's organ*. This structure is said to respond to the removal of the adjacent testis or to the injection of female sex hormones by enlarging to become structurally like an ovary. Occasionally isolated ova have been found within the seminiferous tubules of an otherwise normal testis, suggesting the similar origin and the fundamental similarity of the testis and the ovary.

Finally, attached to the anterior end of the testis of the hibernating frog may be seen finger like *fat bodies* (corpora adiposa) which represent stored nutrition for the long period of hibernation, and for the pre-breeding season when food is scarce. Under the microscope these fat bodies appear as clusters of vacuolated cells, and are not to be confused with the mesorchium. It is believed that they, as well as the gonads, arise from the genital ridges of the early embryo. The fat bodies tend to be reduced immediately after the breeding season, only to be built up again as the time for hibernation approaches.

THE FEMALE REPRODUCTIVE ORGANS

Secondary Sexual Characters

The mature female frog is generally larger than the male of the same age and species, the *Rana pipiens* female measuring from 60 to 110 mm. in length from snout to anus. The sexually mature female has a body length of at least 70 mm. It can be identified by the absence, at any season, of the dark thumb pad; the inability to produce lateral cheek pouches resulting from the croaking reaction; a flabby and distended abdomen; and the presence of peritoneal cilia. These cilia are developed in the female in response to the prior development and secretion of ovarian hormones.

The Ovaries

There is a pair of large flat, irregular multi-lobed ovaries attached to the dorsal body wall by a double-layered extension of the peritoneum known as the *mesovarium*. This peritoneum continues around the entire ovary as the *theca externa*. Each lobe of the ovary is hollow and its cavity is continuous with the other 7 to 12 lobes. The ovaries of the female are found in the same relative position as the testes of the male but the peritoneum extends from the dorso-mesial wall rather than from the kidneys, as in the male.

The size of the ovary varies with the seasons more than does the size of the testis. From late summer until the spring breeding season the paired ovaries will fill the body cavity and will often distend the body wall. They may contain from 2,000 (*Rana pipiens*) to as many as 20,000 ova (*Rana catesbiana*), each measuring about 1.75 mm. in diameter (*Rana pipiens*). The mature ova are highly pigmented on the surface of the animal pole, so that the ovary has a speckled appearance of black pigment and white yolk, representing the animal and the vegetal hemispheres of the ova.

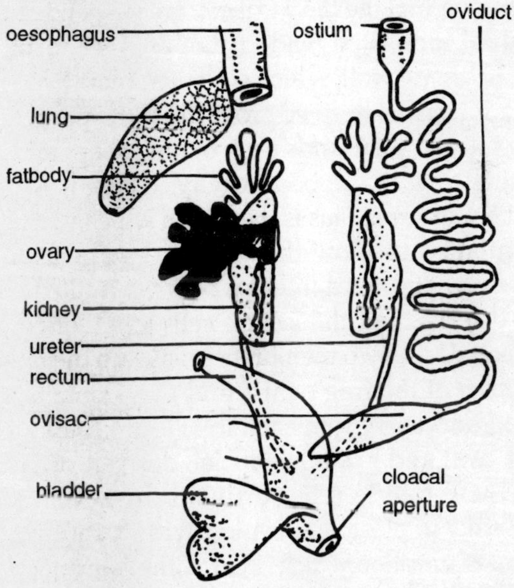

Fig. 1.6. Frog. Urinogenital system of female.

All reproductive organs undergo cyclic changes. There is no appreciable change in the size of the ovary during hibernation, nor is there any observable cytological change in the ova. However, if a female is forced to retain her ova beyond the normal breeding period by isolating her from males or by keeping her in a warm environment and without food, the ova will begin to deteriorate (cytolize) within the ovary. Immediately after the spring breeding season, when the female discharges thousands of *mature ova*, the remaining ovary with its *oogonia* is so small that it is sometimes difficult to locate. There is no pigment in the tissue of the ovary (in the stroma or in the immature ova), and each growing *oocyte* appears as a small white sphere of protoplasm contained within its individual *follicle sac*.

Fig. 1.7. Frog. T.S. of ovary.

The history of the ovary shows that within its outer peritoneal covering, the *theca externa*, are suspended thousands of individual sacs, each made up of another membrane, the *theca interna* or cyst wall, which contains smooth muscle fibers. This theca interna is derived from the retro-peritoneal tissue. The theca interna surrounds each ovum except for the limited area bulging toward the body cavity, where it is covered by only the theca externa. This is the region which will be ruptured during ovulation to allow the ovum to escape its follicle into the *body cavity*. The theca interna, plus the limited covering of the theca externa, and the follicle cells together comprise the *ovarian follicle*. These two membranes make up the rather limited *ovarian stroma* of the frog ovary, and they contain both blood vessels and nerves. Within each follicle are found *follicle cells*, with their oval and granular nuclei, derived originally from oögonia. These follicle cells surround the developing oocyte and are found in close association with it throughout those processes of maturation which occur within the follicle. Enclosed within the follicle cells, and closely applied to each mature ovum, is the non-cellular and transparent *vitelline membrane*, probably derived from both the ovum and the follicle cells. This membrane is developed and applied to the ovum during the maturation process so that it is not seen around the earlier or younger oögonia. Since the bulk of the ovum is yolk, this membrane is appropriately called the *vitelline membranes*. It is sometimes designated as the primary egg membranes. After the ovum is fertilized this membrane becomes separated from the ovum and the space between is then known as the *perivitelline space,* filled with a fluid. The fluid may be derived from the ovum which would show compensatory shrinkage. As the oocyte matures and enlarges, the follicle cells and membranes are so stretched and flattened that they are not easily distinguished.

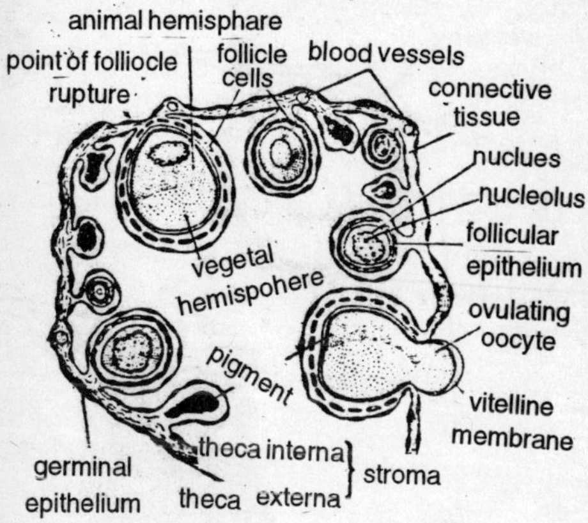

Fig. 1.7

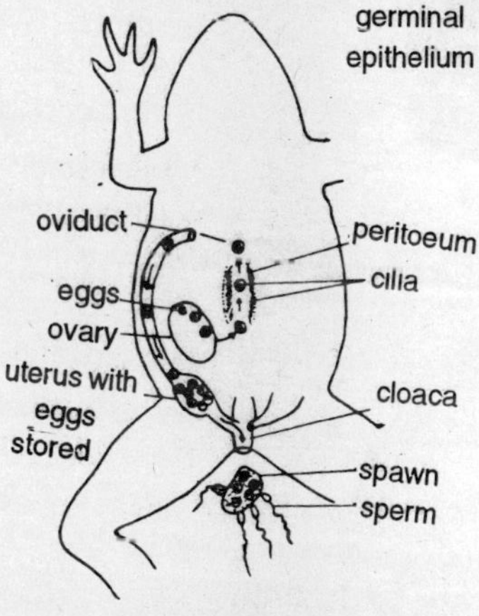

Fig. 1.8. Diagram showing egg transport in frog

The ovum will mature in any of a variety of positions within its follicle, the exact position probably depending upon the maximum blood supply. As one examines an ovary the ovum the will be seen in all possible positions, some with the *animal hemisphere* and others with the *vegetal hemisphere* toward the theca externa and body cavity. It is believed that the most vascular side of the follicle wall will tend to produce the animal hemisphere of the ovum and hence give it its fundamental symmetry and polarity.

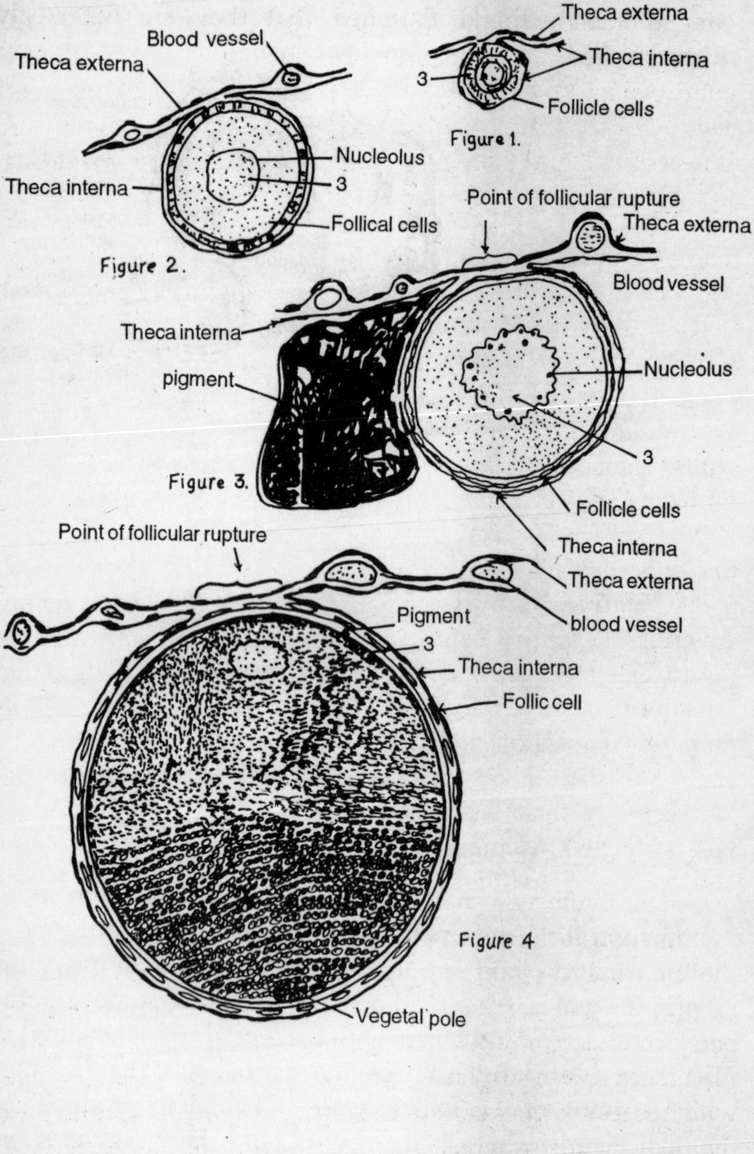

Fig. 1.9. Growing Oöcytes of the frog.

The frog's egg is of *mesolecithal type* having a moderate amount of yolk. The eggs are *telolecithal* having yolk at its lower *vegetal pole*. There is a thin outer layer of *cytoplasm*, more concentrated toward the animal hemisphere and in the vicinity of the *germinal vesicle* or immature nucleus. Surrounding the entire ovum is a non-living surface coat, also containing pigment. This *pigment* is presumably a metabolic by-product. This coat is necessary for retaining the shape of the ovum and in aiding in the morphogenetic processes of cleavage and gastrulation.

The Body Cavity and the Oviducts.

There is a pain of oviducts suspended from the dorsal body wall by a double fold of *peritoneum*. Its anterior end is found between the heart and the lateral peritoneum, at the apex of the liver lobe. At this anterior end is a slit-like infundibulum of *ostium tuba* with ciliated and highly elastic walls. The body cavity of the female is almost entirely lined with cilia, each cilium having its effective beat or stroke in the general direction of one of the ostia. These cilia are produced in response to an ovarian hormone and therefore are regarded as secondary sex characters. They are found on the peritoneum covering the entire body cavity, on the liver, and on the pericardial membrane. There are no cilia on the lungs, the intestines, or the surface of the kidneys except in the ciliated peristomial (peritoneal) funnels which lead into the blood sinuses of the kidneys. The abundant supply of cilia of the female means that ova ovulated from any surface of the ovary will be carried by constant ciliary currents anteriorly toward and into one or another of the ostia.

As soon as the ovum leaves the ovary it is nude except for the non-living, transparent, and closely applied *vitelline membrane*. Thus far it has been impossible to fertilize these body cavity *ovum* and have them develop. When they are placed in a sperm suspension some will show surface markings which resemble very closely the normal cleavage spindles and the cleavage furrows but none have developed as embryos as yet. These body cavity ova are often quite disported, due to the fact that the ovulation process involves a rupture of the follicle and forcing out of the ovum from a very muscular follicle. The ovum is literally squeezed from the follicle, through a small aperture. The process looks like an Amoeba crawling through an inadequate hole. *Ovulation* which is the emergence of mature one from ovary to the body cavity takes several minutes at laboratory temperatures, and is not accompanied by hemorrhage. By the time the ovum reaches the ostium (within 2 hours), as the result of ciliary propulsion, it is again spherical.

Ciliary currents alone force the ovum into the ostium and oviduct. The ostial opening is very elastic is very elastic and does not respond to the respiratory or heart activity. The ova are simply forced into the ostium, from all angles, stretching its mouth open to accept the ovum. As soon as the ovum enters the oviduct and begins to acquire an albuminous (mucin-jelly) covering, it becomes fertilizable. One can remove such an ovum from the oviduct by pipette or by cutting the oviduct 1 inch or more from the ostium, and can fertilize such an egg in a normal sperm suspension. The physical (or chemical) change which occur between the time the ovum is in the body cavity and the time it is removed from the oviduct, which make it fertilizable, are not yet understood.

As soon as the ovum is propelled through the oviduct by ciliary currents, it receives coatings of *albumen* (jelly). The initial coat is thin but of heavy consistency, and is applied closely to the ovum. The ovum is spiraled down the oviduct by its ciliated lining so that the application of the jelly covering is quite uniform. There are, in all, three distinct layers of jelly, the outermost one being much the greater in thickness but the less viscous. The intermediate layer is of a thin and more fluid consistency. There is hyperactivity of the glandular elements of the oviduct just before the normal breeding season, or after anterior pituitary hormone stimulation, so that the duct is enlarged several times over that of the oviduct of the hibernating female.

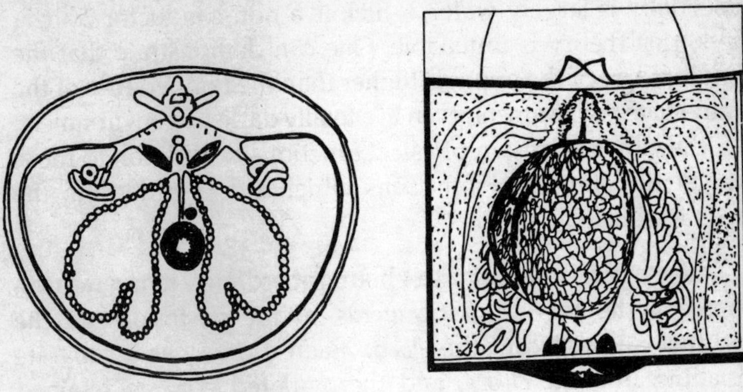

Fig. 1.10 Distribution of coelomic cilia within the body cavity of the female frog. (*Left*) Schematic section through the level of the ovaries. (*Right*) Schematic drawing of the open body cavity. The cilia in the body cavity of the female develop in response to the elaboration of an ovarian hormone, and function in propelling the eggs to the two ostia.

The presence of the jelly layers on the oviducal or the uterine egg is not readily apparent because it requires water before it reaches its maximum thickness. Ova sectioned within the oviduct show the jelly as a transparent coating just outside the vitelline membrane. As soon as the ovum reaches the water, however, imbibition swells the jelly until its thickness becomes greater than the diameter of the ovum.

The function of the jelly is to protect the ovum against injury, against ingestion by larger organisms, and from fungus and other infections. Equally important, however, is the evidence that this jelly helps the ovum to retain its metabolically derived heat so that the jelly can be said to act as an insulator against heat loss. *Bernard* and *Batuschek* (1891) showed that the greater the wave length of light the less heat passed through the jelly around the frog's egg, in comparison with an equivalent amount of water and under similar conditions. Originally, and erroneously, the jelly was thought to act as a lens which would concentrate the heat rays of the sun onto the ovum, but the since jelly is largely water, which is a non-conductor of heat rays, this theory is untenable. One can demonstrate that the temperature of the ovum is higher than the temperature of the immediate environment, even in a totally darkened environment. So, the jelly has certain physical functions in addition to those as yet undetermined functions which aid in rendering the eggs fertilizable.

The egg takes about 2 to 4 hours, at ordinary temperatures, to reach the highly elastic *uterus*, at the posterior end the oviduct and adjacent the *cloaca*. Each uterus has a separate opening into the *cloaca*, and the ovulated ova are retained within this sac until, during amplexus (sexual embrace by the male), they are expelled into the water and are fertilized by the male. Generally the ova are not retained within the uterus for more than a day or so. There may be quite a few hours between the time of appearance of the first and the last ova in the uteri.

Oögenesis

The process farming ova is called *oogenesis*. During the process the ova develop from oogonia which divide repeatedly. These pre-maturation germ cells divide by mitosis many times and then come to rest, during which process there is growth of some of them without nuclear division. These become ova while those that fail to grow become follicle cells. However, there are pre-prophase changes of the nucleus of the prospective ovum comparable to the pre-prophase changes in spermatogenesis. The majority of oögonia, therefore, never mature into ova, but become follicle cells.

The process of maturation involves contributions from the nucleus and the cytoplasm. *First*, chromatin nucleoli aid in the synthesis of yolk, and *second*, the breakdown of the germinal vesicle allows an intermingling of the nuclear and the cytoplasmic components. Only a small portion of the germinal vesicle is involved in the maturation spindle so that it may be at this time that the nucleus exerts its initial influence on the cytoplasm. All cytoplasmic differentiations must be initiated at a time when the hereditary influences of the nucleus are so intermingled with it.

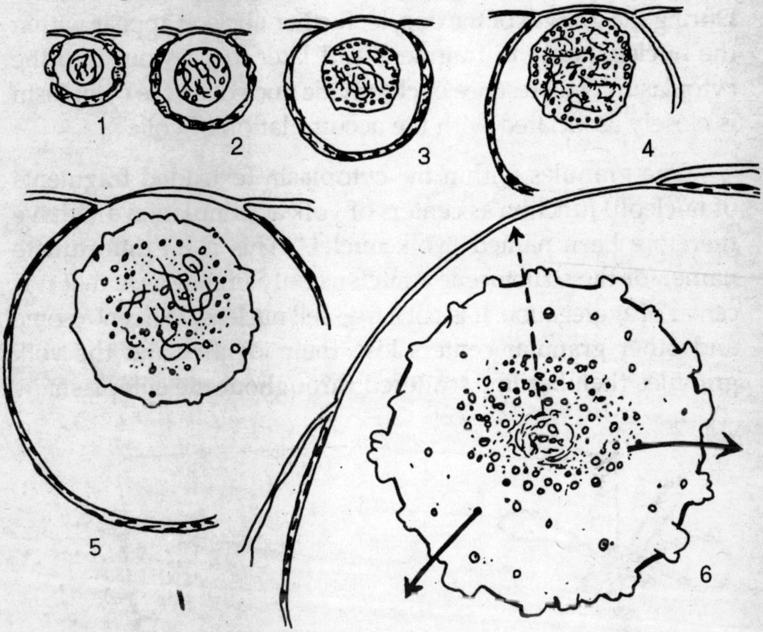

Fig. 1.11. Normal nuclear growth cycle and its role in vitellogenesis.

Growth of the developing oocytes is achieved largely by the accumulation of *yolk*. As soon as growth begins the cell no longer divides by mitosis and is known as an *oöcyte* rather than an oögonium. The growth process is aided by the *centrosome*, which is found to one side of the *nucleus*, and around which gather the granules or *yolk platelets*.The *chromatin* filaments become achromatic and the *nucleoli* increase in number, by fragmentation, and become more chromatic. Many of the nucleoli, which are concentrations of nucleo-protein, pass through the nuclear membrane into the surrounding cytoplasm during this period. It is not clear whether this occurs through further fragmentation of the nucleoli into particles of microscopic or sub-microscopic size, and then their ejection through the nuclear membrane. It may occur by the loss of identity (and chromatic properties) by possible chemical change and subsequent diffusion of the liquid form through the membrane to be resynthesized on the cytoplasmic side of the membrane. During the growth of the oöcyte, further nucleoli appear within the nucleus, only to fragment and later to pass out into the cytoplasm. The presence of chromatic nucleoli in the cytoplasm is closely associated with the accumulation of yolk.

The granules within the cytoplasm (extruded fragments of nucleoli) function as centers of yolk accumulation and have therefore been named "yolk nuclei." This is an unfortunate name, for the structure is a nucleus only in the sense that it is center of aggregation. It is not a true cell nucleus. The centrosome and other granular centers lose their identity and the yolk granules then become scattered throughout the cytoplasm.

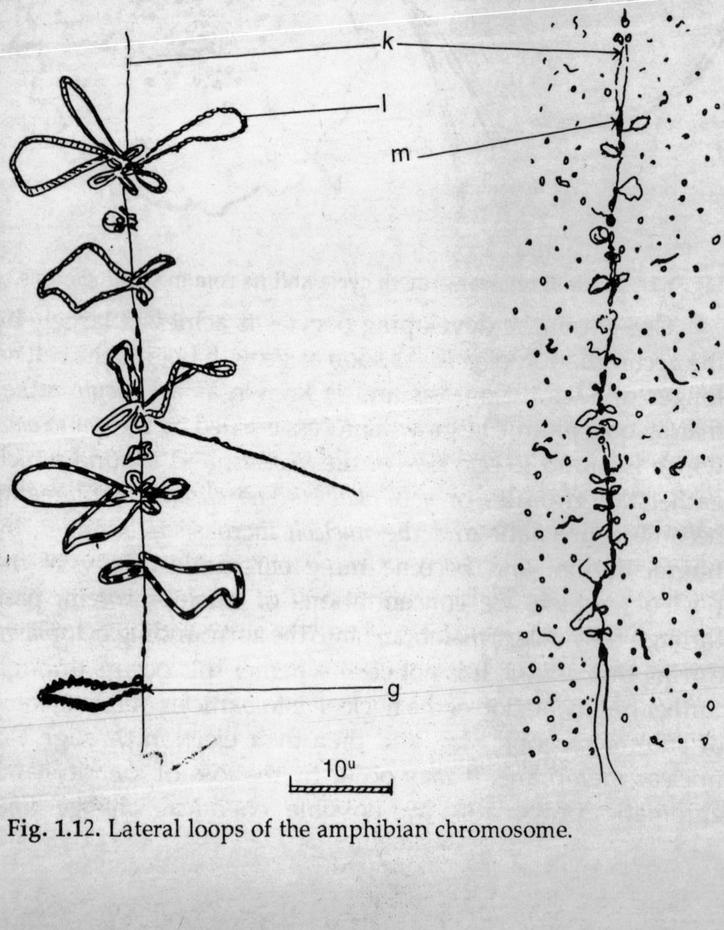

Fig. 1.12. Lateral loops of the amphibian chromosome.

The source of all yolk for the growing ova is originally the digested food of the female. This nutrition is carried to the ovary by way of the blood system and conveyed to the nurse or follicle cells and thence to the oocyte. The yolk is at first aggregated around yolk nuclei, then concentrated to one side of the nucleus. Finally, it assumes a ring shape around the nucleus between an inner and an outer zone of cytoplasm. Subsequently the nucleus is pushed to one side by the ever-increasing mass of yolk so that eventually there is an axial gradient of concentration of oval *yolk platelets* from one side of the egg to the other. The smaller platelets are found in the vicinity of the nucleus, in the animals hemisphere. The larger platelets are located toward the vegetal hemisphere. There is an increase averaging from 200 to 700 per cent in the total lipoid substance, neutral fat, total fatty acids, total cholesterol, ester cholesterol, free cholesterol, and phospholipin content of the ovaries of *Rana pipiens* occurring during the production and growth of ova. The primary oöcyte may show a slight flattening of the surface directly above the region of the nucleus.

These growth changes and the unequal distribution of pigment, yolk, and cytoplasm are the first indications of *polarity* or a gradient system within the ovum. When the polarity is well-established, the cytoplasm, the superficial melanin or black pigment, and the nucleus are all at the animal hemisphere. The light coloured yolk is more concentrated toward the vegetal pole. The eggs is then regarded as a *telolecithal* egg. During this phase of egg maturation there is a drain on the metabolism of the frog which requires an excess of food intake because the materials for the growth of ovum must be synthesized from nutritional elements received from the vascular system of the female. For *Rana pipiens* this period of most active feeding comes during the summer when the natural foods, insects, worms, etc., are the most abundant.

During the growth of the oocyte in general there are important changes occurring within the nucleus (germinal vesicle) of the ovum. Thirteen pairs of *chromosomes* may be

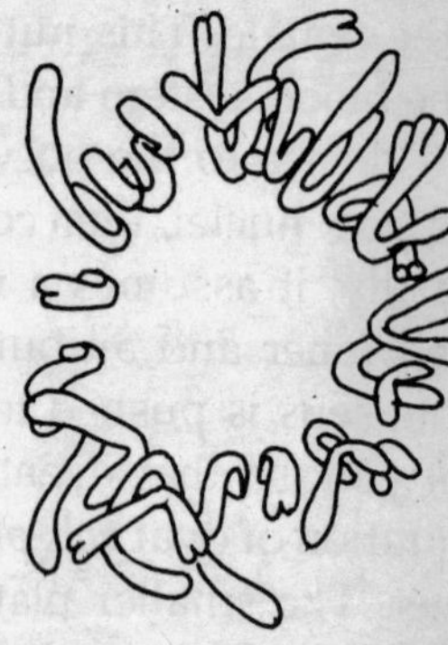

Fig. 1.13. Diplid metaphase chromosomes from the tail fin of a 15-day-old *Rana pipiens* tadpole.

seen in synizesis (contraction), converging toward the centrosome at the "yolk nucleus" stage. A little later the nuclear membrane develops sac-like bulges, the *nucleoli* are scattered, and there is a colloidal chromosome core which almost fills the entire nucleus. The chromosome themselves are small and almost invisible. When the *primary oocyte* is about half its ultimate size, there appear definite sacs on the nuclear surface. The fragmented nucleoli are located at the periphery of the lobulated nuclear membrane, and the chromosome frames have become relatively large. The chromosomes, by this time, have reached their maximum length and possess large lateral loops. Finally, in the fully grown nucleus of the primary oocyte the nuclear sacs are very prominent, and the nucleoli appear in clusters in the center of the ovum surrounding the *chromosome frame*. This frame is a gel structure which give rise to the *first maturation spindle*, containing 13 pairs of slightly contracted chromosomes.

Before the time of hibernation the ova that are to be ovulated for the next spring are in the fully grown primary oocyte stage, having their full complement of yolk, cytoplasm, and pigment. Externally more than one-half of the ovum appears densely black, due to surface pigment granules, while the rest is creamy white. The nucleus is prepared for the maturation divisions. Such an ovum measures about 1.75 mm. in diameter. The surface layer of the amphibian ovum is formed before fertilization and it is definitely not hyaline, as it is in some Invertebrate eggs. It contains many small yolk grains and irregular accumulations of spherical, black pigment granules. With each cleavage, subsequent to fertilization, this superficial coat is divided between the blastomeres, being an integral part of the living cell. There is no clear-cut demarcation between this surface coat and the inner cytoplasm and yolk. It is believed that these growth changes of the egg are under the influence of the basophilic cells of the anterior pituitary gland, which cells are greater in number at this time than at any other.

During the growth period the vitelline membrane appears on the surface of the oocyte as a thin, transparent, non-living, and closely adherent membrane. It is formed presumably by a secretion from the egg itself, aided by the surrounding follicle cells. It appears to be similar in all respects to the membrane of the same name found around the eggs of all vertebrates.

Ovulation and Maturation

Ovulation, or the liberation of the ovum from the ovary, is brought about by a sex-stimulating hormone from the anterior pituitary gland. Just before and during the normal spring breeding period there is a temporary increase in the relative number of acidophilic cells in the anterior pituitary. It is believed that this is not coincidental but a causal factor in sex behaviour in the frog. However, until such time as extracts of specific cell types can be made this will be difficult to prove conclusively. Attempts on the part of the male to achieve amplexus are resisted by the female not sexually stimulated. However, such a female can be made to accept the male by injecting the female with whole anterior pituitary glands from other frogs. it is very probable that environmental factors such as light, temperature, and food may act through the endocrine system to prepare the frogs for breeding when they reach the swampy marshes in the early spring, after protracted hibernation. It must be pointed out, however, that there are frogs in essentially

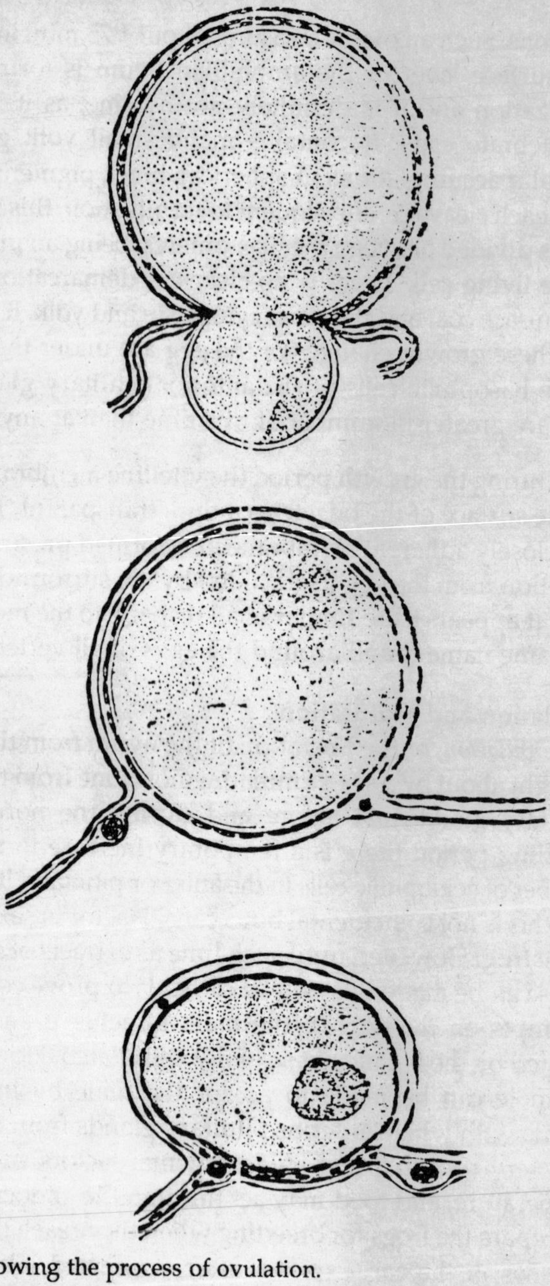

Fig. 1.14. Showing the process of ovulation.

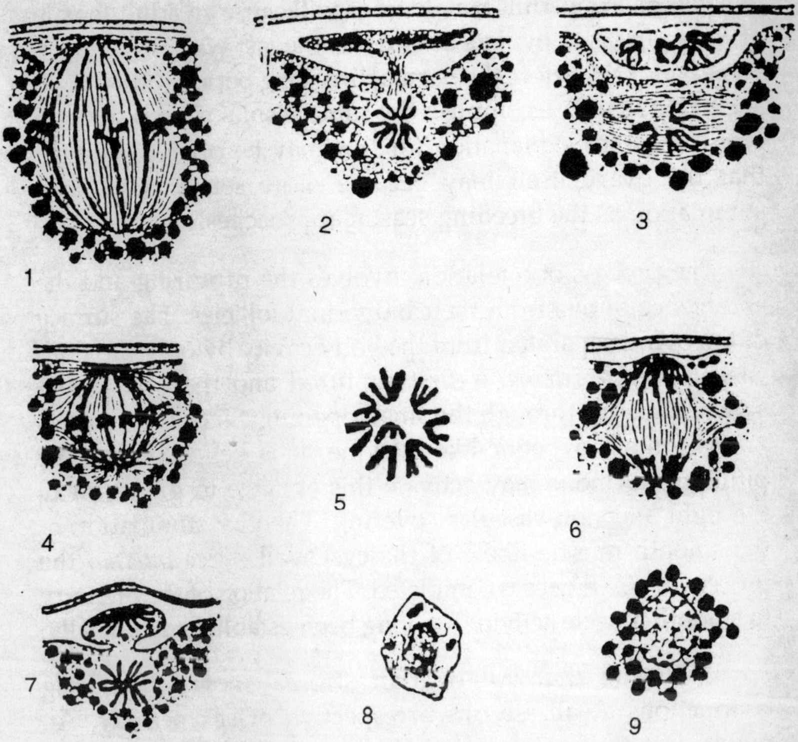

Fig. 1.15. The maturation divisions in the female (Axolot). (1) First polar spindle with heterotypic chromosomes. (2) Extrusion of first polar body. (3) Appearance of second polar spindle. Longitudinal division of chromosomes in egg and in first polar body. (4) Second polar spindle radial. Homoeotypic chromosomes on equator (metaphase). (5) Polar view of the same. (6) Anaphase. (7) Extrusion of second polar body. (8) Second polar body with resting nucleus. (9) Female pronucleus in resting condition, closely surrounded by yolk granules.

the same environments which breed in July (*Rana clamitans*) and August (*Rana catesbiana*), so that the causal factors appears to be either complex or possibly different for different species. The pituitaries of the hibernating frogs do contain the sex-stimulating factor, but apparently to a lesser degree than the glands of frogs approaching the breeding season. The injection of 6 glands from adult female frogs will cause an adult female of *Rana pipiens* to ovulate as early as the last week in August, some 8 months before the normal breeding period. One or two such glands will accomplish the same results if used early in April. Another explanation for this may be offered, namely that the ovary itself may become more sensitive to such stimulation as the breeding season approaches.

The process of ovulation involves the rupturing and the emergence of ova from their individual follicles. The surface of the ovum separated from the body cavity by only the non-vascular *theca externa*, is first ruptured and then the ovum slowly emerges through the small opening. Since the ovum is know contain a peptic-like enzyme, it is believed that the pituitary hormone may activate this enzyme to digest away the tight and non-vascular covering. Then by stimulation of the smooth muscle fibers of the cyst wall *(theca interna)* the process of emergence is completed. The relation of the pituitary to smooth muscle activity has long been established clinically.

It is true that the ovarian stroma show undulating contractions at all seasons, irrespective of sex activity. An ovum will emerge at any time from a surgically ruptured follicle. If an ovulating female is etherized and the body cavity is opened, the ovary may be removed and placed in amphibian Ringer's solution and the ovulation process may be observed directly. This will go on for several hours after all connections with the nerve and blood supply are cut off. From the initial rupture of the theca externa until the ovum drops free into the body cavity there is a lapse of from 4 to 10 minutes at ordinary laboratory temperatures.

The *first maturation division* occurs at the time of ovulation. The heterotypic chromosomes are placed on the spindle of the amphiaster whose axis is at right angles to the ovum surface. Movement of the chromosomes is identical with that found in ordinary mitosis. The outermost group of telophase chromosome are pinched off, with a small amount of cytoplasm and no yolk, to comprise the *first polar body*. The innermost telophasic group of chromosomes remain within a clear area of the ovum as the nuclear mass of the secondary oocyte. These changes occur as the ovum leaves the ovary and before it reaches the oviduct. Possibly the same forces which bring about follicular rupture also influence this maturation process.

The *second maturation division* begins without any intermediate rest period for the chromosomes, at about the time the ovum enters the oviduct. There may be a variation in time up to 2 hours for ova to reach the ostium, depending upon the region the body cavity into which they are liberated. Thus the stage of maturation of different ova within the oviduct may vary considerably. There is a longitudinal division of the chromosomes of the ovum which are lined up in metaphase on the *second maturation spindle*, the axis of which is at right angles to the ovum surface. Since the spindle is primarily protoplasmic, and is made up in part of fibers the space occupied by the spindle will be free of yolk. Since it is peripherally placed, and represents a slight inner movement after the elimination of the first polar body, the surface layer of the ovum is slightly de-pigmented just above the spindle region. This situation is exaggerated in aged ova a relatively large de-pigmented area of the cortex appearing toward the center of the animal hemisphere.

Maturation is not completed until or unless the ovum is activated by sperm or stimulated by parthenogenetic means. However, every ovum reaching the uterus is in methaphase of the second maturation division, awaiting the stimulus of activation to complete the elimination of the *second polar body*.

2

Courtship and Mating in Amphibians

It is in spring that the average person is most conscious of the amphibian. The Spring Peeper, *Hyla crucifer*, is considered by many to be more reliable harbinger of warm weather than is the robin. One old saying is that there will be but three more freeze-ups after the peepers are heard calling. This is akin to the superstition that if a ground hog sees his shadow on the second of February, there will be six more weeks of winter. Though the peepers are weather prophets far superior to our finest meteorologists, unfortunately they are not so expert that they can forecast just how many more frosts will occur before it is safe to set out the tomato plants.

A great many amphibians breed in the spring, but by no means all of them do. Some breed in the summer, some in the fall, some in the winter. However, in the North we are most apt to notice the breeding activity in the spring. Spring must be interpreted very loosely, for sometimes the peepers and Wood Frogs are calling before March 21st even as far north as New England. The nights are often cold and frosty, and you may wonder why these little creatures have come out of hibernation so early.

We do not begin to recognize all the conditions that cause the amphibians to emerge from hibernation and to commence courtship and mating, any more than we understand all the reasons behind bird migration. But though there is much to be learned on this subject, it is interesting helpful to discuss a few of the factors that have an important bearing on the matter.

The prime factor governing the amphibian's emergence form hibernation is the weather. If the ground is frozen solid, the animal hibernating beneath the frost line would not be able to break through the frozen crust, even should he so desire, which he does not. For those that hibernate in the detritus of pond bottoms, ice on top prevents their escape. But the amphibian is a cold-blooded creature, whose internal temperature is nearly the same as his external surroundings, and very cold surroundings make him sluggish and incapable of the violent movement necessary to burrow out of hibernation. Therefore, until the temperature of the earth or pond water rises to a certain level, the amphibian remains in the dormant condition known as hibernation, where heartbeat and respiration are slowed to the minimum.

All amphibians are extremely sensitive to temperature changes, and notice the slightest gradations that are imperceptible to us. Furthermore, each amphibian has a temperature at which he must live for optimum comfort and activity. This varies from species to species, and helps to explain why the Spring Peeper emerges form hibernation and begins to breed earlier than the toad.

The temperature of the water is important not only because it must be warm enough to arouse those sleeping beneath it to activity but also because it must be warm enough so that any eggs laid in it will not be frozen and killed.

Humidity also plays a tremendous part in determining courtship and mating time. You will always find that the peak of breeding activities in all species occurs on evenings of high humidity — the choruses of peepers, toads, treefrogs, and frogs are at their loudest on rainy overcast days and especially nights. Naturally, the humidity of the air has more influence on those amphibians who are more terrestrial than on those who are partially or completely aquatic. But it affects them all to some extent. Perhaps there is considerable truth in the fisherman's adage that fish bite better on a rainy or dull day. Certainly it would be true if one fished for amphibians.

The weather does not, of course, completely govern courtship and mating. It is also regulated by the animal's ductless glands—especially the anterior lobe of the pituitary gland. The secretions of this gland stimulate the sexual glands, which in turn rouse the animal to mate.

We usually think of ponds and lakes as being the place where amphibians congregate to mate and lay their eggs. It is true that the great majority of all amphibians do deposit their eggs in water. The salamanders may choose a quiet pond, as do the Newt and many Mole Salamanders. They may prefer the swift-running water of a brook, as do many Lungless Salamanders. Or they may select a moist place in the woods or in the mud. Doubtless, this dry-land laying is a surprise to many. And no wonder, for it is far more difficult to find a few eggs under a stone or log in the woods than it is to see the thousands upon thousands clustered in various parts of the pond. It may be even more of a surprise to learn that some salamanders lay their eggs in trees.

Frogs and toads lay in much the same places as do the salamanders. There are pond layers, river layers, and even tree layers, though none of our native frogs lays in trees. With the frogs and toads, courtship is always conducted very close to the place where mating and egg laying will occur. How do they court one another? Let us take a specific example, the Spring Peeper.

The peepers hibernate in the woods, beneath the earth. When the ground becomes thawed, they emerge from their winter burrows and head for ponds and swamps. On dull afternoons and in the evenings, the male sits on the bank of the pond and utters his sweet birdlike call. He takes air into his lungs and closes his mouth and nostrils. He sends the air into the vocal sacs located in the throat region. They balloon out into a glistening bubble that is almost as large as the body of the peeper. The air is sent back and forth between lungs and vocal sacs, causing the vocal chords to vibrate. An enormous sound for so small a creature emerges: *Preep, peep, peep* ... he sings sweetly. Other males join in, and soon there is a deafening chorus of peepers all calling to the females. A female approaches one of the singing males. She is silent, for only the males sing in the breeding season. But the male senses her approach. He turns toward her, climbs on her back, and embraces her, placing his arms just behind hers in her "armpits." The couple are now ready to enter the pond and lay the eggs. Courtship in frogs and toads consists mainly of the male's calls. The female approaches the calling male. He may be so intent on his singing that she may even have to touch him to make him aware of her. Female toads frequently nudge males, though in other species of frogs and toads, movement in the vicinity of a calling male is sufficient to make the male cease his calling and turn to embrace the moving object.

Since the peak of the breeding activities are carried out at night, sight plays no part in the recognition of a mate. Sound and hearing and feeling are the things that count. The females are guided to the males by their calls. The males hear or feel a nearby female. "But," you may ask quite reasonably, "how do the males know they are embracing a female and not another male? And how can they be sure that they have a mate of their own species, and not a female of another species, if they cannot see her?"

First of all, the males distinguish the females by their silence. Should a male embrace another male — and this happens quite frequently — the male who is thus treated squeaks or croaks in protest, and struggles to free himself or such unwanted attentions. The female, however, is completely silent during the breeding season. Some female frogs can call at other times of the year, but when the breeding season comes they are mute.

Second, different species breed at different times. They are not all in or around the ponds at once. However, there are several separate species there at the same time. The males of like species tend to congregate in groups apart from others. Since each species has its own distinct song, the females naturally go toward the sound of their own kind calling.

Should the male embrace a female of a different species, however, he can tell by feel that she is not of his own kind. She must be the correct size — larger than he, and swollen with unlaid eggs, but not so large that she is of a different species. The way she moves when he embraces her may also tell him whether or not she is of his own kind.

It sometimes happens, however, that mistakes are made in the excitement of the moment. Toads, whose nuptial embrace is strong, have been known to grasp a fish and not let go until the fish was crushed to death. This is rare, but it is probably due to the fact that there weren't enough females to go around and that the urge to reproduce was so strong that the male seized the first object that nudged him.

About the only romantic part of the courtship of frogs and toads from a human standpoint are their songs. These may be as sweet as are the calls of birds, and many people mistake the trill of the American Toad and some of the treefrogs for birds. They may be plaintive, as is the lamblike bleating of the Narrow-mouth Toad. They may be insectlike, as are the calls of the Cricket Frogs. They may be squirrel-like, or like the noise of riveters; they may sound like a banjo, like a snore, or like a foghorn. It all depends upon the species, and there is but one male frog in all the world that, so far as we know, lacks a voice with which to call to his mate. This curious exception, which we shall discuss more fully later, is the Tailed Frog, *Ascapus truei*, found in parts of the northwestern United States.

When we look at courtship in the salamanders, we perceive an altogether different picture. In the first place, because the salamanders lack vocal chords they are silent. Not for them

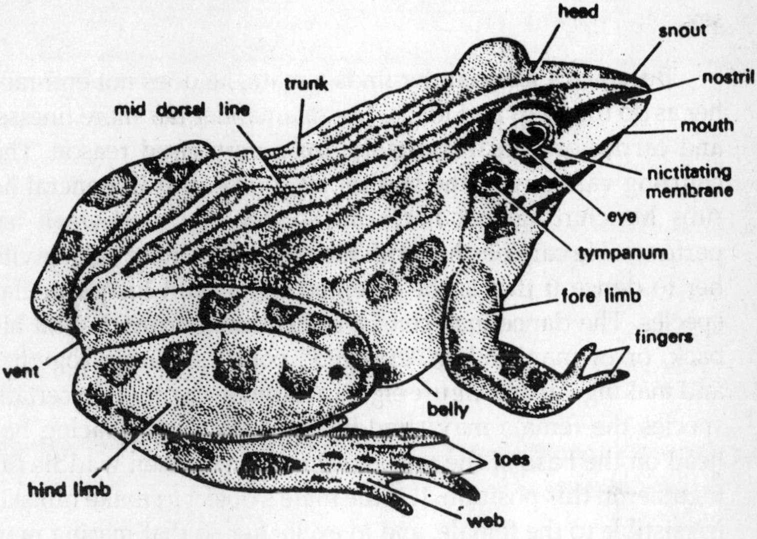

Fig. 2.1. Rana tigrana.

are the pleasures of serenading a lady love on a warm humid night. Neither have they any ears; they "hear" through their front legs or, in certain aquatic species, the lower jaw, which are especially adapted to receive vibrations from the ground. So even if the males could sing, neither the singer nor the one sung to could hear the sounds. They are like the frogs and toads in that sight plays no part in the recognition of a mate. How then do the salamanders find their mates, and recognize them? So far as we know, scent is the main factor in recognition of sex and species. Females and males smell differently, and each species evidently has its own distinctive fragrance. We humans with our dull sense of smell cannot detect most of the odorus that attract the salamanders to one another. The salamanders have their own built-in perfume factories. The hedonic glands, as these "factories" are aptly named, are located mainly at the base of the tail and on the underside of the head on the males. Though the females lack any readily differentiated glands, their skin secretions evidently possess the odour that

others that delay. The Mudpuppy, *Necturus maculosus maculosus*, generally mates in the fall, but does not lay her eggs until the following May or June. The same is true of the Red-backed Salamander, *Plethodon cinereus cinereus*, a terrestrial species.

As we pointed out, most salamanders fertilize the eggs internally. However, there are a few exceptions. In this country there are two families of salamanders that fertilize the eggs externally. The first of these families is the Hellbenders, or Cryptobranchidae. They are completely aquatic salamanders. The male digs out a large nest in the water under a rock. The females enter, and as they lay their eggs the male sheds his sperm over them.

The other family of salamanders, the Sirens, or Sirenidae, so far as is known, fertilizes its eggs externally also. All anatomical evidence points to this conclusion. However, no one has yet observed the Siren breeding. It may seem amazing that no one has yet seen and accurately described these vital statistics of an animal found, not in darkest Africa, but in the United States. On the other hand, anyone who has ever tried to observe animals in the wild will realize how long it takes to obtain data on even the most inconsequential phase of their life cycles.

Mating in frogs and toads is an different matter altogether. Most frogs and toads fertilize their eggs externally. The male frog mounts the female, puts his arms around her, and grasps her firmly either just in back of her arms or in her groin. "Firmly" is a mild word for this nuptial embrace. If you catch a mated pair of toads. you will find it almost impossible to separate them, so tight is the male's grasp and so strong is the clasping reflex. However, it does not seem to bother the female in the least. Possibly, she enjoys this bear hug. In any case, if the pair mate on land they soon enter the water.

The female does not, however, always lay her eggs immediately. She may lay them anywhere from 3 to 26 days after pairing, the couple remaining in amplexus all during this time. What is it that governs when the eggs shall be deposited? Is it some physiological factor? Do some species require prolonged clasping in order to lay? Or does the length of time between coupling and laying depend mainly on the weather? Perhaps all of these have some bearing on the matter.

As the female lays her eggs, the male sheds his sperm on top of them. After all the eggs are laid, the female is considerably slimmer, for she may have had as many as 20,000 eggs inside her body. As soon as she no longer feels so pleasingly plum, the male releases her. Does he return to the bank and recommence his calling in the hope of attracting another female? According to one authority, frogs and toads fertilize the egg complement of but one female each year, but this may be wrong, at least for some species. It is an extremely difficult thing to prove either way.

What of the female who finds no mate? Does she lay her unfertilized eggs just the same? Evidently not. As with many unmated birds, the eggs of the female are gradually resorbed into the body. A few may emerge from the oviduct, but they are never deposited in the characteristic mass of the species as they are when the male clasps the female.

There are several outstanding and interesting exceptions to the way most frogs and toads mate. We shall discuss three of them.

The first example is the primitive aquatic Surinam Toad known as *Pipa pipa*. This large flat-bodied creature lives in the ponds of South America. The mating call is a rapid clicking. This metallic sound is not produced by air vibrating the vocal chords, as in all other frogs, but by the cartilaginous disks of two bones "cracking" as the bones are moved. In much the same manner, some people cn "crack" the joints of their fingers, jaws, or knees.

Clasping before laying is prolonged, and lasts at least 24 hours and sometimes longer. During this period the skin on the female's back becomes swollen and puffy. For many years it was believed that *pipa* had a protrusile oviduct that was everted when laying began. However, it has recently been proved that this is not true. When the couple are ready to lay, the pair roll over and, while they are in an upside-down position, three to five eggs emerge and are fertilized. The pair then right themselves. The entire roll and righting take about 11 to 14 seconds, and the pair is upside-down for about one second of this entire time. When the female is ready to lay more egg, the pair once more roll over, until, after many such turns, the entire complement of from 40 to 114 eggs are laid. The eggs are pressed into the spongy skin on the female's back by the male's ventral surfaces; the first eggs are placed near the anus, later ones successively farther forward toward the female's shoulders. One day after deposition, the eggs are half buried in the skin of the female, and by the tenth day the top membranes of the eggs are level with the skin of the back.

Our own Tailed Frog, *Ascapus truei* is voiceless. Since he does not seem to be able to hear, and since he spends most of his time in swift-running mountain streams, a voice would be of little use to him anyway. He possesses something far more valuable, namely, a tail. Actually, it is not a real tail, though it certainly looks like one. It is an organ for copulation. The male crawls along the bottom of the stream searching for a mate. When he finds a female, he embraces her and then inserts his "tail" into an extension of her cloaca, and thus her eggs are fertilized internally.

The Tailed Frog alone possesses an intromittent organ because of his habitat. External fertilization of eggs would be precarious, to say the least, in the fast-flowing water where the Tailed Frog makes his home. No sooner would the sperm be shed than the water would carry it downstream before it could possibly fertilize all the eggs of the female. So in order that the species may continue to reproduce. Nature has provided the Tailed Frog with a "tail" so that he may fertilize his eggs internally.

There are two African toads with the jawbreaking names of *Nectophrynoides vivipara* and *Nectophrynoides tornieri.* These toads also fertilize their eggs internally. But the mystifying thing in their case is how they go about it. They have no copulatory organ as has the Tailed Frog. Scientists have long been puzzled as to how the sperm is introduced into the female's body. Do they, like the female salamanders, pick up a spermatophore which the male deposits. There is no indication that they do. But scientists know that the eggs *are* fertilized internally because both species give birth to fully transformed young. The entire larval period is spent in the mother's body, and when the young are born they are miniatures of their parents. *Nectophrynoides vivipara* and *Nectophrynoides tornieri* and the only two known species of frogs and toads that are ovoviviparous, though there are several salamanders that are normally ovoviviparous and several that are occasionally so when existing conditions, such as extreme cold or high altitude, make egg laying impractical.

Let us look at some of the external differences between the sexes. With the sole exception of the Tailed Frog, these differences are secondary sex characteristics in frogs and toads. They are to a frog what a beard is to a man. With the Tailed Frog, the male is easily distinguished from the female by his longer "tail." Her "tail" is an extension of her cloaca, and is much shorter and more blunt than is his.

With the rest of the frogs and toads, we find great variety in secondary sex characteristics. About the only secondary sex characteristic that most male frogs and toads possess in common are their vocal sacs — and even here there are one or two exceptions that we shall examine in a movement. These vocal sacs are of different sizes and shapes, according to the species. They may be in throat region and when in use swell out to a glistening bubble, as is true in the peepers and many toads. They may swell out in kidney-shaped masses on the side of the head, as in the Leopard Frog. Or there may be no localized swelling, but a general enlargement of the throat, as in the Bullfrog.

Vocal sacs might seem absolutely necessary to the male frog, as without them he would be unable to call to the females in the breeding season. But that is evidently not so, for some of our frogs, such as the California Toad, *Bufo boreas halophilus*, and a few others, lack vocal sacs altogether. The Tailed Frog lacks them also, but he is believed to be silent. However, the California Toad is known to have a voice — a trill, similar in character to that of our northeastern American Toad, *Bufo americanus americanus*, but lower pitched. Neither does the sound carry so well, but whether this is because of the lack of vocal sacs, as one eminent scientist believes, or because lower-pitched sounds do not seem to carry so well as those that are shrill, has not been proved.

The vocal sacs, even though they are usually present, are not very helpful to us when we pick up a frog and want to know whether or not he is a male. Unless we see his vocal sacs swelled up, we would not even know he had them. The females of some species lack vocal sacs, and in others they are greatly reduced in size. And, as we have said, the females are silent during the breeding season, though at other times of the year they may make calls similar to the males of their kind, but not so loud or so resonant.

Size is a factor in sex recognition of frogs and toads. The female is a general rule, larger than the male. It is just as well that she is, because she must carry him about on her back during mating and egg laying. In the water, though the additional weight does not matter so much, the female nevertheless does the swimming for both. Many males keep their hind legs drawn up on the female's back, while others allow their legs to float out behind. In a few species, if the couple is alarmed, the male may try to aid the female by using his feet in swimming motion, but their efficiency is impaired by his position on top of her. Many, however, like the American

Toad, make no attempt at all to help the female in her race to escape. It seems curious to us that the male refuses to release his hold on the female and that each does not go its separate way when threatened with danger. But such is not their habit. Some species separate more easily than others, but none does it voluntarily, and often it is almost impossible to force the mated pair apart without injury, so strong is the embrace of the male.

Whereas the female is usually larger in over-all size, the male may have much larger and heavier forelegs. Perhaps, he has need for powerful arms the better to grasp the female. The male toad, *Bufo boreas*, found in the Northwest and in California, has enlarged forelimbs. His are insignificant, however, compared with a South American frog. *Leptodactylus ocellatus*, whose arms are fully three times the size of those of the female!

In some species of frogs, the eardrum or tympanum, of the male is considerably larger than the eye whereas in the female it is the same size. A male Bullfrog can easily be distinguished from a female by this characteristic. Why the male should need a larger ear than the female has never ben explained. You would think that if either of them should have a larger one, it would be the female that would have most use for it. And why, of all species, should the Bullfrog, with his tremendous foghorn cry that can be heard from a distance of half mile or more, need a larger ear. The answer probably lies in the Bullfrog's evolution. At one time male perhaps did need larger ears, and though they are no larger useful they have not as yet been discarded.

In certain species of frogs and toads, the males have much larger thumbs. These may be swollen at the base or at tip. This enlargement of the thumb gives the male better gripping power. Does it also help to cushion the female's skin from being overly bruised by the male amphibian's strong grasp? We have never heard this suggested as a reason — it would probably be considered rank sentimentality by most scientist — but we can see no reason why it should not be so.

There is also a difference in the shape of the toe webbing of the two sexes in certain species. A male Wood Frog and be distinguished from a female in the breeding season by his convex toe webs. The female's are concave. Of course, you have to catch the frog and spread his hind toes apart in order to see this. It can't be observed from a distance.

Some reptiles and many birds present striking color dimorphism between the sexes. This is not true of frogs and toads. Colour differences, when they exist, are usually confined to the underparts, and especially the throat region. Some males have darker or more brightly coloured throats than the females of the same species.

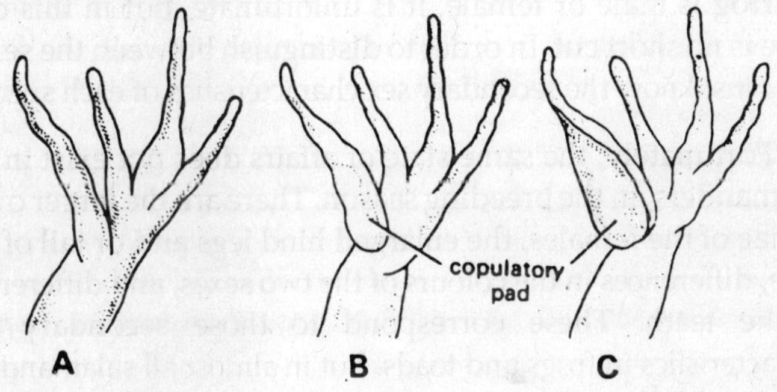

Fig. 2.2. A, Foot of frog showing web, B, Hand of male frog showing nupial or amplexusory pad.

Of all the secondary sex characteristics exhibited by the frogs and toads, none is more startling than those developed by the male Hairy Frog. *Astylosternus robustus*, of Africa. All male frogs and toads need a great deal more oxygen during the breeding season than at any other time. This is due in part to their vociferous calls, and in part to the fact that their metabolic rate is heightened. Most male frogs land toads, therefore, have larger lungs to satisfy the necessary requirements of their bodies for oxygen. But the Hairy Frog is an aquatic frog with very small lungs. Since he is a large fellow, he needs a great deal more oxygen in the breeding season than his diminutive lungs can supply. Therefore, he broadens his skin area at that time, so that he will have a greater surface through which to absorb the extra oxygen needed, by growing long "hair" on his hind legs and lower sides. Of course, they are not really hairs, any more than gills and feathers, which they resemble at times. Scientists call these "hairs" vascular villosities. Be that as it may, they look like hair and they work like gills, enabling the Hairy Frog to obtain the additional oxygen necessary during the breeding season.

All in all, the secondary sex characteristics of frogs and toads are not very useful to us in helping us to tell whether our frog is male or female. It is unfortunate, but in this case there is no short cut. In order to distinguish between the sexes, you must know the secondary sex characteristics of each species.

Fortunately, the same state of affairs does not exist in the salamanders, in the breeding season. There are the larger overall size of the females, the enlarged hind legs and or tail of the male, differences in the colours of the two sexes, and differences in the teeth. These correspond to those secondary sex characteristics in frogs and toads. But in almost all salamanders, there are differences around the vent, especially during the breeding season, though at other times these may be negligible. The male's vent is large, protuberant, and has tiny nipple-like projections. The female's vent is smooth or in folds.

Before we go on to the eggs. Let me urge you to choose a warm humid overcast night in May for an outing. Wear a pair of galoshes, take a strong flashlight, and head for the nearest pond or swamp. You should have an experience that you will not soon forget. Peepers, treefrogs, toads, and frogs will be calling from all sides. You should be able to approach them easily, for they depend mainly on sight to warm them of enemies, and at night they cannot see you. Look in the water. There you will see mated pairs, salamanders and frogs, calling under the water. You will notice species you never knew existed, and species that you rarely if ever catch a glimpses of during the day. You have read how the frogs, toads, and salamanders court and mate but there is nothing that can compare with the thrill of actually witnessing these things with your own eyes and ears. Even if you aren't a keen field naturalist, you will be amazed at all you will discover, and will be very glad that you went. Perhaps you may even be so impressed and enthusiastic that you will want to repeat the experience.

3

The Eggs

The beginning of most life is hidden from us. We are unable to see the egg being fertilized; its development from one-celled organism to an embryo of many cells is screened from our view until birth or hatching. This is true of the shelled eggs of reptiles and birds. They are fertilized within the parent's body and are then surrounded with a shell before laying. The mammalian egg is shell-less with two exceptions; but it is microscopic in size and inaccessible for observation, since it remains inside the mother's body during its development.

If we really want to find out what happens between laying and hatching we must turn to the amphibian egg. The spermatozoa of the amphibian are microscopic in size, so that we are unable to see (with the naked eye) the egg being penetrated by the sperm. But all stages after fertilization are visible to use; furthermore, most of the changes that occur can be seen without a microscope.

To begin with, the eggs are round, though they do not remain so. They may be black or brown on top. The top part is known as the animal pole; it is the section that is alive and will grow. The vegetal pole at the bottom may be white or yellow. It consists of the food on which the egg "feeds" as it develops, for you must remember that all living things; be they animals or that most miraculous of all living entities, a fertilized egg cell, must have nourishment if they are to grow. The living protoplasm of the animal pole is lighter than the non-living deutoplasm of the vegetal pole. Therefore, it always is uppermost. You can easily test this by turning an amphibian egg so that the animal pole faces down. You will notice that very little time elapses before it has rotated so that the protoplasm is again facing up. A few amphibian eggs are yellow or white, and are called unpigmented eggs. But though the separation of animal and vegetal poles is not readily visible in these, it exists nonetheless.

Because the eggs require a certain amount of insulation from shocks land enemies, as they descend through the female's oviduct they are coated with albuminous jelly. Depending upon the way the oviduct is constructed in different species, there may be one or more jelly envelopes, or envelopes surrounding the eggs may be lacking, and the mass of jelly may constitute the sole protection. You can get a clear picture of what an egg with its jelly envelope looks like if you make a small dot and draw a circle around it. The jelly may be clear, or it may be milky white or greenish.

Once the eggs are deposited in the pond, the jelly swells. The capacity of the jelly to absorb water is a decided benefit to both the layer and the egg. It means that the eggs and jelly, when inside the female amphibian, require little space. But one in contact with the water, the jelly swells to twice or three times its original bulk, giving far greater protection to the egg than would be the case if it did not enlarge. Of course, there are instances on record of frogs that delayed too long in laying their eggs. The water content of the body is absorbed by the jelly, which expands rapidly and beyond all bounds. The female then bursts open and dies. Fortunately such circumstances as these occurs seldom, and may be considered freaks of nature.

The eggs vary in size according to the species that lays them. But the size of the adult is no reliable indication of the size of the egg that will be laid. The Tailed Frog is only about 2 inches long. But it lays one of the largest eggs of all our native-frogs. Each vitellus is approximately 5 mm., or roughly $^1/_4$ inch in diameter. Our larges frog, the Bullfrog, which may be as long as 8 inches, lays small eggs of about 1.5 mm. The smallest egg in this country is laid by the Southern Chorus Frog. *Pseudacris nigrita nigrita* and is approximately .7 mm. in diameter. We shall allow you to figure out the equivalent measurement in fractions of an inch. Frankly, it's beyond our mathematical ability, and we are quite content to leave egg measurements in millimeters. The average salamander lays a far larger egg than do the frogs and should you ever wish to watch the development of the amphibian egg, you would be well advised to choose those of a salamander rather than those of a Bullfrog or a toad.

The number of eggs laid also varies greatly, and generally speaking, salamanders lay fewer eggs than do frogs and toads. The Bullfrog and the American Toad lay as many as 20,000. The Rocky Mountain Toad, *Bufo woodhousei woodhousei*, produces upto 25,000 eggs, the largest complement laid by a native salientian. The Robber Frogs of the genus *Syrrhophus* may lay as few as 5 eggs but they are the exception. Most frogs and toads lay eggs in the hundreds or thousands, while salamanders lay in the tens or a hundred. Because of the salamander's more secretive ways, larger egg size, and internal fertilization, they do not need to lay so many eggs to ensure continuance to the species as do the frogs.

The way the eggs are laid also differs with the species. The Red-spotted Newt lays each egg separately, fastening each on the leaf of a water plant and folding the leaf around it thus hiding it very effectively from enemies. The Spring Peeper also lays its eggs individually, but this species merely deposits each egg on the pond bottom. The Bullfrog and the Green Frog both lay theirs in spreading mass on the surface of the water. The Wood Frog and the Spotted Salamander lay theirs in a round or oval mass attached to submerged vegetation. The American Toad and most other toads lay theirs in two long strings of jelly. The Hellbenden that large aquatic salamander, lays its eggs in rosarylike strings, as does the Tailed Frog. The

Spadefoot Toads lay their eggs in bands or cylinders. It is, of course, the jelly that gives the egg mass its shape, for the eggs themselves are round, with the exception of those of the Colorado River Toad who occasionally, though not always, lays wedge-shaped eggs. The form of the jelly is governed in part by the way the oviduct is shaped, and in part by the natural egg-laying movements, which differ in the various species.

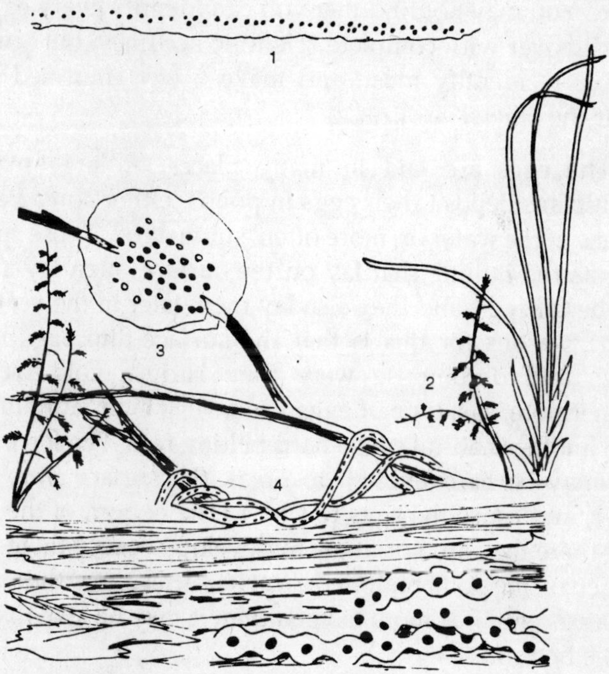

Fig. 3.1 The eggs of some amphibians: 1. The surface mass of the Bullfrog. 2. eggs of the Red-spotted Newt laid individually among plant leaves. 3. The submerged egg mass of the Spotted Salamander. 4. The egg complement in strings laid by the American toad. 5. Eggs in rosary-like strings deposited by the Hellbender.

Eggs of amphibians are identified one from another by their colour, size, number of jelly envelopes, if any; shape of the eggs mass, total complement of eggs, and the places in which they are laid. All of this sounds complicated as do most things scientific at first glance. However, it is not so formidable as it seems, and if you are one who enjoys finding out just exactly what species of amphibian has laid those eggs in the southeast corner of your pond, do not be overawed. Consult one of the two handbooks listed in the Bibliography, and forge ahead. You may not find that you can identify every egg that you discover with complete scientific accuracy, but you will be able to identify most, and make a well-informed guess about the others.

The eggs are laid in many places. A large majority amphibians deposit their eggs in ponds, either spread on the surface of the water or, more often, submerged. Those species, such as the Bulling that lay on the surface often lay a large number of eggs, and they also lay them later in the year. One of the reasons for this is that the surface-film eggs would freeze and decompose were they laid in cold weather. Furthermore, this type of eggs has a very high mortality rate from another natural cause-hard pelting rain. The force of the rain may be hard enough to break the surface film of the water, and cause the eggs to sink to the bottom of the pond. And there they are not likely to develop because there is less oxygen on the pond bottom than at its surface, and the surface-film eggs enquires more oxygen than it may be able to obtain on the bottom.

More salamanders lay in the swiftly running waters of brooks and rivers than do frogs. Undoubtedly, this is primarily because most of them fertilize their eggs internally, whereas the frogs, with the exception of the mountain-stream-dwelling Tailed Frog, are unable to do so. Those salamanders that lay in rivers either scoop out a nest, with its entrance downstream, for protection from the onslaught of the cascading waters, or they attach their eggs, either singly or in a mass, to the underside of a rock.

In many parts of this country, permanent bodies of water are non-existent.Where this situation occurs, you might think that there would be no water-laying amphibians. But amphibians such as the Spadefoots, the Narrow-mouths, and some of the toads must lay in shallow water, and in dry regions they lay in mud puddles that are formed from a rain that may not recur for a year or more. Naturally, the eggs hatch rapidly, and the tadpoles are transformed in a short time into adults.

The habit of laying in the shallow water of mud puddles and syamps is not, however, confined to arid regions. None of our north-eastern toads will deposit their eggs in deep water even though it is available. If you wish to find toad's eggs, the best place to look is a swamp or a mud puddle. We have even found toad's eggs laid in a large mud puddle in the middle of a road. Lacking swamps in your vicinity, go to a small pond. Search around its shallowest edges, for it is there that you will find the toad's strings of jelly.

Very few people who do not make a study of amphibians are aware that some of them deposit their eggs on moist or dry land. This is not particularly surprising, for the chances of finding these land-laid eggs just by accident are slim, hidden as most of them are beneath rocks or logs. As a matter of fact, diligent hunting is generally required to locate them even if you know the most likely spots to search. There are quite a few native salamanders who lay on land. The Marbled Salamander, *Ambystoma opacum*, lays in a moist spot near the water, as does the Dusky Salamander, *Desmognathus fuscus*. The Red-backed Salamander, *Plethodon cinereus*, is woodland salamander, and it lays its eggs in cavities or rotted logs. For some reason the Amphiuma, *Amphiuma means*, though it is an aquatic salamander, also lays its eggs in muddy spots on land. There are many reasons for normally terrestrial amphibians to lay on land, but why a salamander that spends its life in the water should do so is a mystery.

As to frogs, there are fewer of them, and most of them are

not so common as the land-laying salamanders. The land-laying frogs in this country are confined to several species within one family, the Robber Frogs, or Leptodactylidae, of which we will say more in a subsequent chapter.

We have but one native species of amphibian that deposits its eggs in trees. This is the Oak Salamander, *Aneides lugubris lugubris*, of California. It occasionally lays on the ground, but just as frequently lays in trees.

None of our native frogs lays in trees, but there are many foreign frogs that do. All the treefrogs of the family Hylidae that live on the island of Jamaica lay in trees. Perhaps this habit arose because there are very few natural ponds and lakes in Jamaica. The type of vegetation found in Jamaica and parts of South America favoured its development. The Jamaican hylas all deposit their eggs in bromeliads. Bromeliads are relatives of the pineapple, but, unlike it, they are epiphytic; that is, they attach themselves to trees and obtain their sustenance from the air through specially modified aerial roots. The bromeliad leaves from a rosette, just as do those of the pineapple. In the very center of the rosette is a hollow that generally holds a considerable amount of water. It is here that the Jamaican hylas place their eggs, and the tadpoles develop in this water. Even in the driest of seasons, the bromeliad has a store of water in the center of its leaves.

Other foreign frogs, such as the treefrogs of the family Rhacophoridae, found in Asia, deposit their eggs in trees overhanging pools, and upon hatching, the tadpoles fall into the water below and carry on their larval life there. In order to protect the eggs from drying out, the mother first deposits some albuminous jelly without any eggs. This is whipped into a frothy mass by the hind legs of one or both parents until it is as full of air as beaten eggs whites. The jelly-covered eggs are then placed on top and are covered by an additional layer of jelly which is also well beaten. The outside of this nest soon hardens, while the inside liquefies, providing the eggs with a suitable medium in which to develop.

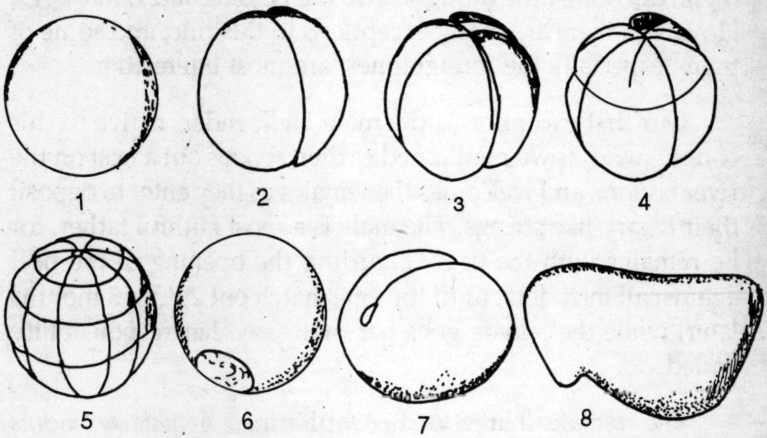

Fig. 3.2. A few stages in the development of a frog's egg: 1. The unsegmented egg. 2. The two-called stage. 3. The four-celled stage. 4. The eight-celled stage, showing the beginnings of cleavage into the sixteen-called stage. 5. The blastula, or many-called stage. 6. The yolk-plug stage. 7. The neural-groove stage. 8. The embryo.

The changes in the amphibian eggs from the moment it is laid and fertilized to the time it hatches are fascinating to watch. This is how most life begins and develops, from the mammal down to the lowly earthworm. If you go to the nearest pond and collect the largest eggs that you can find, you may watch the entire process with the naked eye or with an inexpensive magnifying glass. Suffice it to say here that the eggs changes — divides from one into many cells — lengthens out, and develops, and finally the tadpole or larva wriggles its way out of the confining jelly and hatches. The length of time between laying and hatching varies with the species and the temperature. It may be as short a time as 36 or 58 hours, as in some toads, or as long as 3 or 4 months in some salamanders. In general, salamander eggs are slower in hatching than are those of the frogs and toads.

By and large, amphibians lay their eggs and then desert

them, devoting little thought as to the possible fate of the eggs. However, there are many exceptions to this rule, and some of them, especially the foreign ones, are most interesting.

Our first exception is the male Hellbender, native to this county, who, as we mentioned earlier, scoops out a nest on the river bottom, and welcomes the females as they enter to deposit their rosary-like strings. The male is a most faithful father, for he remains with the eggs, guarding the opening of the nest against all intruders, until the eggs hatch out $2^1/_2$ or 3 months later, while the female goes her own way, her responsibility ended.

The female Three-toed Amphiuma, *Amphiuma means tridactylum*, who lays in the mud, remains with her eggs until they hatch. One wonders why the other subspecies, *Amphiuma means means*, who also lays in the mud, does not remain with hers.

This habit of one or both parents remaining with the eggs laid on land is quite common. We find it true of the Marbled Salamander, the Dusky Salamander, the Oak Salamander, and the Redbacked Salamander. When this habit exists, you may be sure it is necessary, not so much for the protection of the eggs from enemies as from desiccation. The damp body of the parent helps in no small measure to keep the eggs themselves moist. It has also been suggested that the skin secretions and cloacal excretions and / or secretions of the parents may serve as a deterrent to any fungus growths that might destroy the eggs capsule and egg. Recent research indicates, however, that mold develops only on dead eggs or embryos. The role that skin and cloacal secretions and the brooding habit play in increasing the number of young which survive has not been fully explained to date.

When we come to the foreign species who care for their eggs, we see all manner of strange sights. The male Midwife Toad, *Alytes obstetricans*, found in western Europe, is tied down by his parental duties. The female lays her jelly-covered eggs

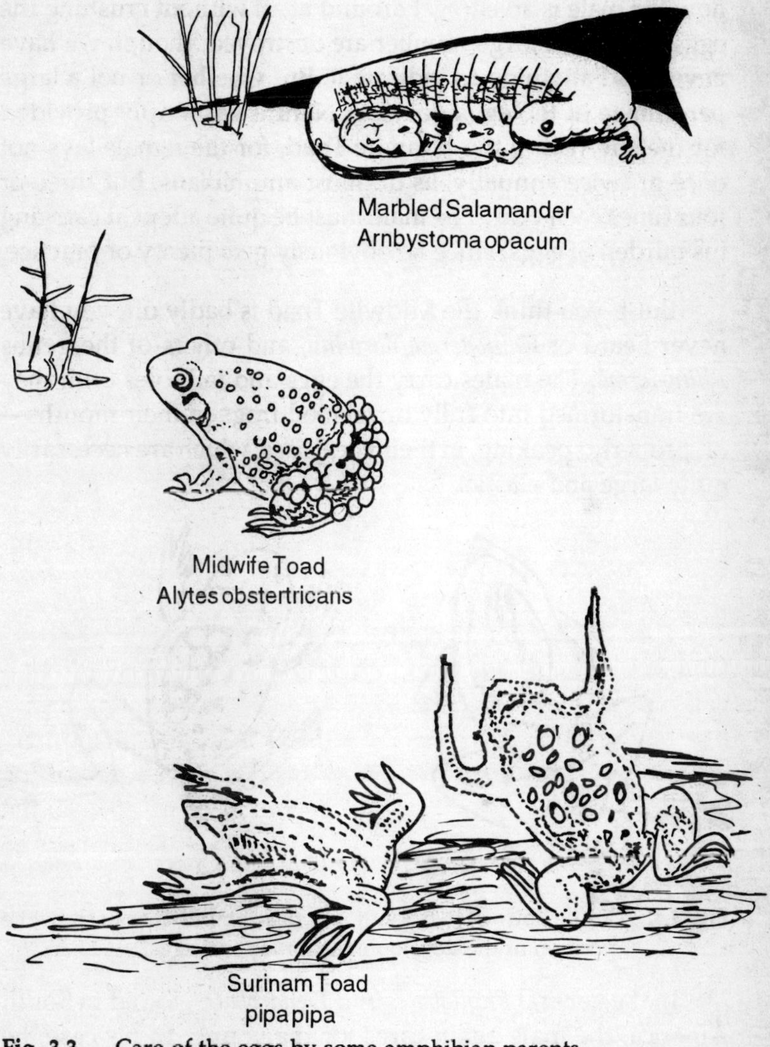

Fig. 3.3. Care of the eggs by some amphibian parents.

in long strings; the male then takes these, wraps them around his hind legs, and retires to a damp cavity. Here he remains during the day, coming out at night to bathe the eggs in dew or in a nearby pond. When the tadpoles are ready to emerge, the male goes to the water and the tadpoles hatch. One wonders how the male is able to get around at all without crushing the eggs. Perhaps a large number are destroyed, though we have never read anything to indicate it. But whether or not a large percentage of the eggs develop, Nature has amply provided for the survival of the Midwife Toad, for the female lays not once or twice annually, as do most amphibians, but three or four times each year. The male must be quite adept at carrying his burden of eggs, since he obviously gets plenty of practice.

But if you think the Midwife Toad is badly off, you have never heard of *Rhinoderma darwinii*, and others of the genus *Rhinoderma*. The males carry the eggs and tadpoles until they are transformed into fully developed frogs in their mouths—or, properly speaking, in their vocal sacs, which are necessarily quite large and elastic.

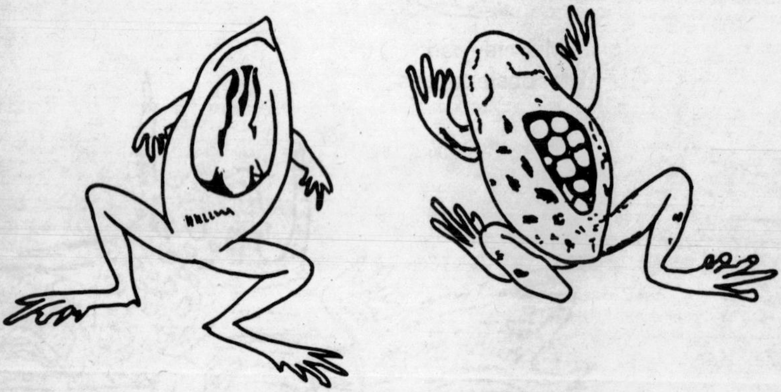

Fig. 3.4. X-ray views of two South American amphibians to show the position of the developing embryos and eggs.

In the general *Phyllobates* and *Dendrobates*, found in South America, the male again cares for the young. In his case, he builds a mud dam to trap water for a suitable site for egg laying. Should this artificially made pool dry up before the tadpoles are transformed, their father takes them for a pickaback ride to a nearby stream where they can complete their development. They, in turn, have special suctorial lips with which they can fasten themselves firmly to his back so that they will not be dislodged on the overland journey. Naturally, if they should fall off their life is soon ended, for though they are able to withstand short trips out of water, they are unable, while tadpoles, to live away from an aquatic medium for any extended period.

4

Development of Frog

The frog is well-known animal, living chiefly in ponds and wamps, though some of its species may be also to exist in damp or shady places away from the water. Eggs of frog are easily obtained and may be studied in the laboratory from the moment of fertilization onward. This is not so easily true of the higher vertebrates. The amphibian embryos, also, are relatively straight, not coiled. This makes description easier. The stages of embryonic development differs in various chordates, yet the chief phases are basically similar in all. The differences are related primarily to the amount of yolk particles present in an egg. The yolk particles provide nourishment to the developing embryo. Here, we shall discuss the entire stages of development right from gametogenesis to adult stage.

Gametogenesis

The sexual reproduction takes place by the fusion of two mature reproductive or germs-cells (*gametes*), which are *sperms* from male and *ova* from female so as to form a single cell, the *fertilized egg* or *zygote*. The gametes are usually developed in parents of different sexes. Male gamete is a sperm cell or *spermatozoon*, the female gamete is an egg-cell or *ovum*. Each gamete is formed by a process, called maturation or *gametogenesis* in gonads. Male gonads are testes, while the female gonads are *ovaries*. The fusion of nuclei of male and female gametes is called *fertilization*. The zygote changes into a mature animal through the process of *embryology i.e., development* and *metamorphosis*.

Gametogenesis is a *process* by which gametes are formed from cells, called germ-cells, a term which includes the gametes themselves. Earliest germ-cells are called *primordial germ cells*, which can be recognised very earlyr in the life of an animal as found in *Ascaris*, where germcells are formed from one of the four cells of the embryo. However, in majority of the other animals, they are distinguished at early stages, before the gonads appear. By mitotic cell-divisions *primordial germcells* give rise to *gonia* or additional germ-cells. The development of sperm cells from the primordial germ-cells is known as *spermatogenesis*, while the development of egg-cells from primordial germ-cells is known as *oogenesis*. In both these processes, the number of chromosomes is reduced to *half*. This reduced number is called *haploid* and, at the time of fertilization, *diploid* number of chromosomes is formed. Thus, the *constant number* of chromosomes characteristic of a given species is maintained.

Spermatogenesis

A typical testis is made up of thousands of cylindrical spermtubules, in each of which numerous sperms develop. The walls of the sperm-tubules consist of spermatogonia or unspecialised germ-cells. For the growth of the testis, the *spermatogona* divide the mitosis so as to give rise to additional spermatogonia. When the sexual maturity is attained, some spermatogonia undergo spermatogenesis, which consists of two *meiotic divisions*, followed by cellular changes resulting in maturation of sperms. Other spermatogonia still continue dividing so as to form additional spermatogonia for spermatogenesis at a later period. During breeding season, testis increases in size and spermatogenesis takes place.

During spermatogenesis, spermatogonia grow into larger cells, called *primary spermatocytes*. Each *primary spermatocyte* divides by first meiotic division so as to form two cells of equal size, called *secondary spermatocytes*. These divide by second meiotic division so as to form four *spermatids*. Secondary spermatocytes are known as *diptoid* (2*x*) *i.e.*, the number of chromosomes is the same as is present in most of the cells of the animal body. But in each spermatid, there is a bit of cytoplasm and the number of chromosomes is haploid (x). Now, no division takes place but spermatid changes into *spermatozoon*. The changes include sprinking of nucleus in size, and formation of head. Most of the cytoplasm is shed off. Some of the *Golgi bodies* aggregate at the anterior end of the sperm so as to form a point, which is said to be helpful for puncturing the cell-membrane of the egg during fertilization. A bit of cytoplasm also develops a *tail*, which helps movements of the sperm. The *mitochondria* lie at the junction of the head and tail so as to form a *middle piece*, which gives energy for the beating of the tail.

The structure of mature spermatozoa is different in different species. Sperm of *Ascaris* is tailless. The sperms of *crabs* and *lobsters* are curious having pointed projections on the head.

Oogenesis

Immature sex-cells of the ovary are called oogonia, which, as a result of mitotic division, from additional *oogonia*. When the individual attains sexual maturity, oogonia develop into large *primary oocytes*. They are much larger in the size than the primary spermatocytes, as *yolk* is found in them. After growth, the primary oocyte divides by first meiotic division; as a result, two daughter-cells are formed. However, these daughter-cells are not of equal size. One daughter-cell is called the *secondary ooayte*, which gets all the cytoplasm and yolk, while the other is called the *polocyte* or *polar body* and is simply a nucleus and may divide into two additional *polocytes*. The secondary oocyte divides by second meiotic division again with an unequal division of cytoplasm so as to form a large *ootid* with yolk and cytoplasm and a small second polocyte. Further cell-division does not take place and ootid changes and the result is formation of a mature *ovum* or *egg*. Polocytes disintegrated and disappear, thus, each primary oocyte forms a single ovum. Polocytes are formed so as to get rid of the excess chromosomes, so that the mature egg gets enough cytoplasm and yolk to survive.

Eggs are of different size, which vary due to the presence of the amount of yolk. Eggs with little or no yolk are small. Human egg is about 0.1 mm in diameter.

Frog's egg is about 2 mm., yolk of hen's egg in about 30 mm. in diameter.

Comparison of Spermatogenesis and Oogenesis

Spermatogenesis	*Oogenesis*
1. Spermatogenesis takes place in testes.	1. Ogenesis takes place in ovaries.
2. As a result, sperms are formed, which are typically much smaller and motile adapted to swim actively to the egg by lashing movements of the tail.	2. Egg is typically large, non-motile and contains yolk, which supplies nourishment to the developing embryo.
3. Secondary spermatocytes divide by second meiotic division to form four spermatids.	3. Secondary oocyte divides by second meiotic division to form a large ootid and a small second polocyte.
4. Four sperms are produced from each primary spermatocyte.	4. Only a single ovum is produced from each primary oocyte.
5. Polocytes are not formed.	5. Polocytes are formed so that mature egg may have haploid number of chromosomes.

EARLY DEVELOPMENT

Fertilization

Fertilization in anurans is normally monospermic. The entrance of the sperm always occurs in the animal hemisphere, though not at the animal pole. The point of penetration is not restricted to any particular meridian of the egg. Following sperm penetration, the perivitelline space widens slightly; the zona radiata lifts away as the fertilization membrane. The egg is now free to rotate within the fertilization membrane, and within an hour or so after fertilized eggs usually rotate so that the pigmented animal hemisphere, which is lighter, is uppermost. The whole sperm normally penetrates the egg. The tail soon disintegrates and the midpiece and sperm nucleus travel along a path (approximately along a radius of the egg) leaving a trail of dark pigment granules. This pigment trial marks the penetration or entrance path of the sperm (Figures 4.1, 4.2). Meanwhile, the penetration of the sperm has stimulated the formation of the second maturation spindle and second polar body, at the animal pole. The egg pronucleus then is positioned below the pole, on the pole axis. Since the sperm penetration path may not be directed precisely toward the egg pronucleus, alteration of course along a new path, may become necessary. This new path is called the copulation path (Figure 4.1). The copulation path is also usually marked by a trail of pigment. Centered 180° from the point of sperm penetration and at the boundary between animal and vegetal hemispheres, the black pigment of the animal hemisphere is withdrawn from a crescent-shaped area, referred to as the gray crescent.

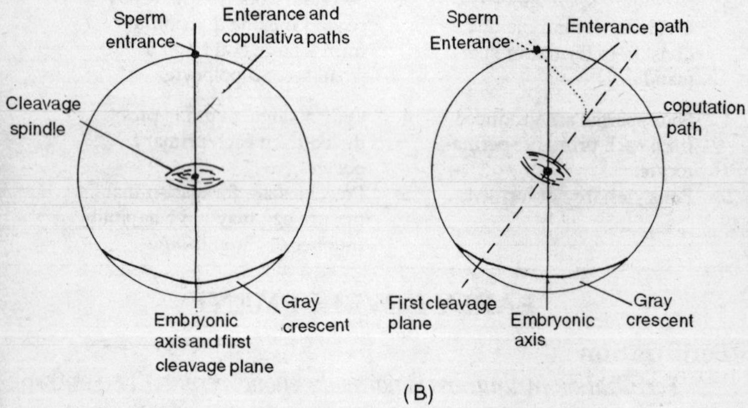

Fig. 4.1. Diagram to illustrate some of the possible relations of the axes in a fertilized frog egg. In both cases, the egg is assumed to be viewed from the animal pole. (A) A egg in which the entrance point plane, the entrance and copulation paths, the gray crescent plane, the first cleavage place and the longitudinal axis of the future embryo all coincide. (B) An egg in which the entrance path and the copulation path are not in the same straight line. Hence, the gray crescent plane and the longitudinal axis of the future embryo fail to coincide with the entrance path, copulation path, or the first cleavage plane.

Ultrastructural changes. At the electron microcope level, sperm penetration is followed, after two or three minutes, by the breakdown of cortical granules. The granules open and empty into the perivitelline space in a manner essentially similar to that observed in sea urchins. About ten minutes after sperm penetration, a system of microvilli appears at the animal hemisphere. It has been suggested that this change may be associated with the displacement of the egg cortex leading to the formation of the gray crescent.

The male pronucleus moves toward the female pronucleus along the copulation path already described. The midpiece (centriole) initiates the formation of the division apparatus, or spindle, which orients at right angles to the copulation path. At syngamy, the pronuclei fuse and the chromosomes line up on the first cleaveage spindle.

Polyspermy. Polyspermy is abnormal in anura, resulting in abnormal chromosome distribution and development when it occurs. An interesting observation has been made on dispermic eggs of frogs. Since the point of sperm penetration determines the position of the gray crescent; the question immediately arises as to the position of the gray crescent in eggs which have been penetrated by two sperm. The answer is the gray crescent appears opposite a point midway between the two points of entrance. The egg seems to average the position, such that both sperm have equal say about the position of the future posterior dorsal side of the embryo.

Physiological polyspermy in urodeles. This condition is one where more than one sperm normally penetrates the egg. It is found in tailed amphibia and in reptiles and birds as well. Still, only one sperm nucleus participates with the egg nucleus in the formation of the fusion nucleus of the zygote. In salamander eggs containing three of the five sperm, the sperm nucleus closest to the egg pronucleus following the second polar body division fuses with the egg nucleus; the remaining (supernumerary sperm degenerate. Constriction of the egg with a fine hair loop into two compartments, one containing the fusion nucleus, and the other the supernumerary sperm, shows that the supernumerary sperm in the compartment with the fusion nucleus degenerate first: those in the other compartment survive longer and in direct relation to the size of the bridge between the two halves of the egg. Those closest to the constriction degenerate first and the wider the constriction, the more rapid the degeneration. The constriction can be drawn completely through the egg, leading to survival of the extra sperm and cleavage in the portion containing the extra sperm. Evidently, a diffusible substance emanates from the fusion nucleus which represses the extra sperm and prevents polyspermy. An alternative hypothesis also fits the observations: Perhaps the fusion nucleus uses up something required for continued life of the extra sperm, leading to their regression. The relation of the effect to the cross-sectional area of the cytoplasmic bridge connecting the two halves would imply a diffusion phenomenon, since diffusion rate is proportional to the cross-sectional area of a diffusion gradient.

FERTILIZATION AND EGG SYMMETRY

Fertilization

Fertilization is external and only one spermatozoon penetrates the ovum. A sequence of events set in to complete the whole process of fertilization.

(*i*) *The penetration of the sperm and the formation of fertilization membrane:* The egg is laid at the secondary oocyte stage at a time when the second maturation division has already reached half-way. Spermatic fluid is shed almost simuitaneously with oviposition and no time is lost in bringing about fertilization. *Monospermy* is a rule in mesolecithal or moderately telolecithal eggs, although several sperms pierce the outer jelly, but usually one of them is slightly in advance of its fellows and thus arrives first at the surface of the egg itself. The moment it has started to enter, cortical granules burst open into the perivitelline space and their contents become adhere to the vitelline membrane which now becomes thick to form the *fertilization membrane.* This membrane checks the entry of other sperms into the egg. Polyspermy is thus an abnormal phenomenon in the frog.

The entrance of the sperm always occurs in the animal hemisphere of the egg, and usually about 40 degree from the pole. The sperm breaks through the egg-membranes with the help of an enzyme, *hyaluronidase.* Further movements of the sperm and its way inside the ovum is probably not an active part of the sperm. The ovum forms a small crator-like projection, the *cone of reception* or of *fertilization,* which engulfs the sperm by a complex process resembling to phagocytosis.

(ii) The perivitelline space and rotation of egg: The penetration of ovum by the sperm has caused some cortical reactions which give up certain amount of its fluid. This fluid collect between the vitelline membrane and the egg plasma membrane and the space containing this fluid is as usual termed the *perivitelline space.* Its formation releases the egg from the grip of its coverings, permits the egg to rotate freely within them. So that the heavier vegetal hemisphere comes to lie downward and the darker and lighter animal hemisphere comes to be upward. Such rotation is called *"rotation orientation."*

(iii)Penetration and copulation path : In case of frog, the whole spermatozoon enters the ovum in the animal hemisphere. It usually takes a minute or two to get entirely inside. The tail then disintegrates; the head and middle piece travel steadily carrying a trail of black pigment behind it which results an area opposite the sperm entrance point, to have a slightly decreased pigment content. This lighter pigmented area is called the *gray crescent.* The straight path followed by the head and middle piece inside the egg is called the *penetration* or *entrance path.* But in case of an excentrically placed female pronucleus, the path followed by the male pronucleus is first straight, then inclined at an angle. The inclined path is called the *copulation path,* and like the penetration path, it is marked

by a trail of pigment. But when the female pronucleus is centrally placed and the sperm nucleus does not change its path, then the *penetration* and the *copulation path* remain the same. As the head and mid-piece of the sperm move along penetration path, the usual rotation of these parts occurs, thus placing the mid-piece in the lead.

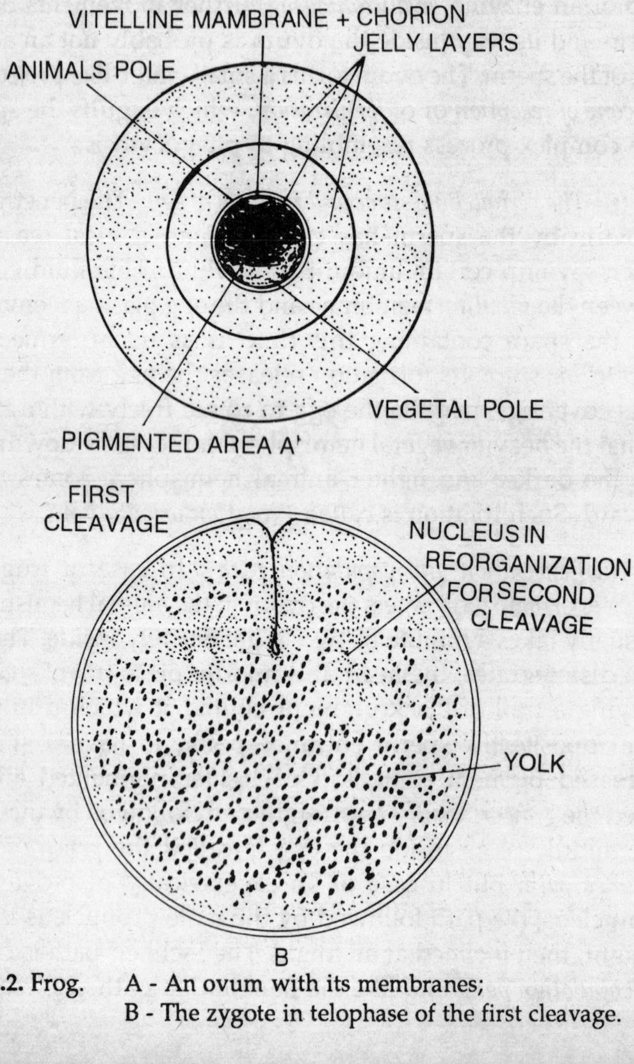

Fig. 4.2. Frog. A - An ovum with its membranes;
B - The zygote in telophase of the first cleavage.

(iv) Completion of second maturation division : Meanwhile the stimulus of the entrance of the sperm has incited the completion of second maturation division of the egg nucleus which was postponed in the mataphase second. After throwing off the second polar body, the egg nucleus withdraws from the surface of the ovum, usually to a position in the egg axis, and becomes vesicular to be called as *egg pronucleus.*

(v) Amphimixis: The sperm nucleus also becomes vesicular and is called *sperm pronucleus.* The sperm pronucleus along with its centriole moves towards the egg pronucleus and ultimately both pronuclei fuse together to form *zygotic nucleus.*

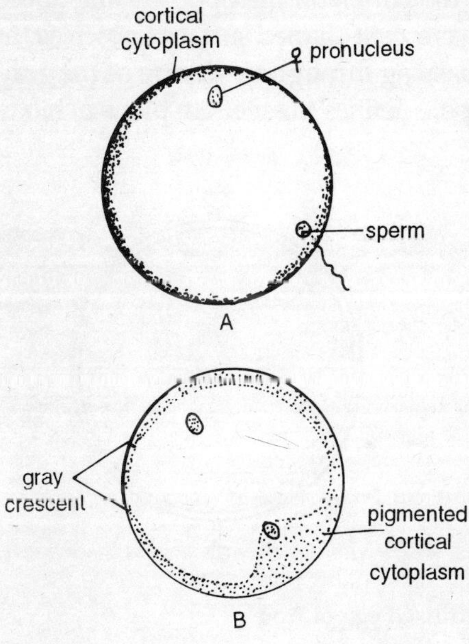

Fig. 4.3. Entry of sperm and formation of gray crescent in the egg of frog.

Egg Symmetrization

The process of bilateral symmetrization of an unfertilized radially symmetrical egg occurs during fertilization and remains intact during the entire embryogenic development. Therefore, the bilateral symmetry of the adult animal originates during fertilization process.

During fertilization, certain very significant changes occur in the egg cortex to establish the bilateral symmetry. There is a temporary shifting of the pigments resulting from a raising and concentration of the cortical layer toward the animal pole, but the components soon returning to their original place. Thereafter, a displacement of the egg contents brought about by a rotation of the cortical layer around the underlying layers, with relative concentration of pigmented layer towards a particular meridian. This is accompanied by a corresponding thinning of the cortex on the opposite site of the egg, thereby producing a crescent-shaped surface known as the *gray crescent*. The plane passing through the centre of the gray crescent and the animal pole defines the median plane of bilateral symmetry

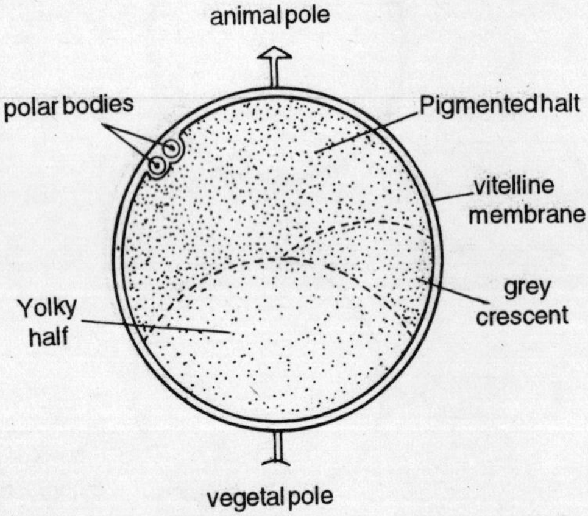

Fig. 4.4. A fertilized egg of frog

in each case, i.e., it coincides with the embryonic axis, and is the only plane which separates the egg into two equivalent parts, each containing half of the crescent material. In frog, the gray crescent represents material which is no longer covered by the pigmented cortical layer, and it is gray because it lacks reflective yolk platelets. Thus, due to slight rotation of pigmented cortex around the endoplasmic mass, the median axis of the bilateral symmetry is established and becomes irreversibly fixed. Such a cortical movement which fixes the median axis of the bilateral symmetrs is called the *"rotation of symmetrization."*

The cortical pigment of the frog's egg not makes visible the shift the cortex relative to the underlying endoplasmic layers but permits determination of the path taken by the sperm head within the egg cytoplasm. As the sperm nucleus and centriole move inward towards female pronucleus from the surface, a train of pigment granules is drawn along. This is the *sperm track,* and it serves as a record of both point of penetration and path taken. It has been found that the sperm track always occurs in the median plane of bilateral symmetry and the point of entrance of the spermatozoon determines the future position of the gray crescent and, therefore, of the axis of bilateral symmetry, because the point of sperm entry and the centre of gray crescent always remain on opposite sides of the egg. This is accomplished by the sperms causing the descent of pigment on one side, the future *ventral side,* the side on which it enters and an ascent of pigment on the other side, the future *dorsal side* of the egg.

Artifical Parthenogenesis

Puncture is the method of choice. Despite many efforts of different investigators to bring about development of the frog egg with various chemical solutions, little more than a few imperfect cleavage furrows have been produced. In 1910, a procedure was adopted which proved successful. The eggs are punctured with a fine needle of glass or platinum. Clean needles do not work satisfactorily. Greatest success was always obtained when the needle had been dipped in frog blood prior to puncturing the egg. Later studies pointed to a role of the blood-clotting apparatus in furthering cell division. Eggs punctured by needles dipped in anti-coagulants, such as heparin, along with whole blood were not successful activated.

Position of the gray crescent. It has been shown in the section on fertilization that the position of the gray crescent is a function of the point of sperm penetration. It would therefore, be reasonable to expect that the localized puncture would simulate the entrance of the sperm and that the gray crescent would bear the same relation to the puncture as it bore to the point of sperm penetration. This is not the case; the gray crescent appears at random after activation by needle puncture and bears no relationship at all to the point of stimulation. It would seem that either the needle fails to simulate the penetration and copulation paths of the sperm accurately, resulting in a loss of correlation with sperm penetration and symmetry, or that the egg has a built-in symmetry of its own, which is overridden by the sperm, but not by the puncture. It has been suggested that needle puncture is too rapid of properly simulate sperm entry and that experiments using a micromanipulator to provide control of the rate of progress of the needle into the cytoplasm are in order. These experiments have not as yet been done, and the question of the point of appearance of the gray crescent during artificial activation remains open.

Regulation to diploidy in frogs. Parthenogenesis produces few viable tadpoles and fewer metamorphosed frogs. In studies of chromosome number, it was noted that the majority of the frogs produced by the puncture method of activation were diploid. Investigations were launched to discover the means of regulation.

The fertilization frog egg normally gives off the second polar body after sperm penetration and prior to DNA replication and cleavage. From studies on *Urechis* it might be expected that artificially activated frog eggs would retain the second polar body following activation. In one study only 13 of 559 activated eggs failed to show a second polar body. Clearly, polar body retention is not the major means of regulation to diploidy. In the majority of cases systematically studied, the diploid individuals showed a cleavage delay of one cleavage division. This leads to the conclusion that regulation in frogs is by *endomitosis* (nuclear division without cytoplasmic division), following complete polar body divisions. Endomitosis results in delay of first cleavage until the time the second cleavage is normally expected. Such cleavage delay has been found to be a reliable indicator of diploidy.

Sex of parthenogenetic. Sex in frogs in influenced by such environmental factors as temperature extremes and length of time the unfertilized eggs have been kept inside the female cloaca before activation or fertilization. In addition, some races of frogs may be female digametic (producing X- and Y-bearing gametes). If these factors are controlled and the species used is male digametic, the sex of the parthenogenetic offspring is preponderantly female (fewer than 5 per cent males). The exceptional males may arise from elimination of one X chromosome (giving rise to an XO individual, who might be male). The mechanism for production of males under these conditions is not known and further speculation is not warranted at this time.

Practical uses of artificial activation. Many experiments require an activated egg cytoplasm, without the female nucleus. These eggs can be produced by artificial activation (puncture method), followed by microsurgical removal of the second polar spindle, or by ultraviolet irradiation to inactivate the egg chromosomes. The technique is used most frequently in nuclear transplant studies, where activated enucleate eggs are prepared to receive nuclear transplants derived from larval or adult tissue or from foreign species.

CLEAVAGE AND BLASTULATION

The cleavage of a fertilized and telotecithal frog egg is of *holoblastic* and *unequal* type. The *first cleavage plane* is *meridional.* This cleavage furrow first appears near the animal pole and progressively extend towards the vegetal pole of the egg. It cuts the egg through its median animal—vegetal polar axis and results in two equal sized blastomeres. The *second set* of divisions is like wise holoblastic and equal, and the planes are *merimional,* but at the right angles to the first plane. But, with

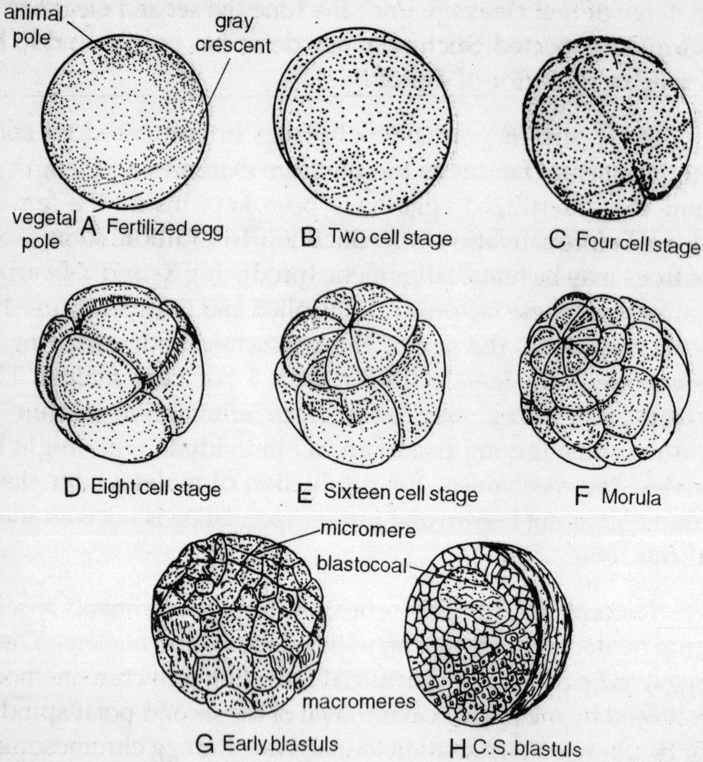

Fig. 4.5. Cleavage and blastulation of frog.

the *third set* of divisions, the unequally distributed yolk makes its influence felt. The mitotic spindles orient parallel to the polar axis, and displaced as they are toward the animal pole. The four blastomeres not only cleave *latitudinally*, but unequally. The eight cell stage, therefore, consists of four small-sized, yolk-rich, vegetal cells, the *micromeres* and four large-sized, yolk poor, animals cells the *macromeres*. By so doing, thus third set of cleavage is set up a differential. The *fourth set* of cleavage planes are *meridional holoblastic*, but, *unequal*. They divide yolk-poor micromeres more rapidly, then yolk-rich macromeres. The macromeres divide rapidly, due to the fate that they are free from the burden of cleaving through a yolk-dense region. The macromeres divide slowly, since their cytoplasm is now loaded with yolk throughout. Yolk platelets have the double effect of conferring physical inertia on the dividing mass and diluting the active cytoplasm with metabolically inactive inclusions. The fourth set of cleavages, thus producing a 16-cell stage.

The blastomeres in the early cleavage stages tend to assume a spherical shape like that of the egg before cleavage. Their mutual pressure flattens the surfaces of the blastomeres in contact with each other but, free surfaces of each blastomere remain spherical, unless these outer surfaces are also compressed by the vitelline membrane. At this stage, the whole embryo acquires a characteristic appearance reminiscent of a mulberry and so, is called *morula*.

It is about the fourth and fifth cleavage stages that a small space, the *blastocoel*, appears within the blastomeres of morula. Actually, this blastocoelic space first may be represented just by narrow cervices between the blastomeres of morula, but, which gradually increases as the cleavage goes on. With the more rapid multiplication of the animal blastomeres (micromeres), the blostocoel shifts more and more toward the animal pole. Meanwhile, the blastocoel becomes infiltered by water and albuminous fluid secreted by the surrounding blastomeres. As cleavage proceeds, the adhension of the blastomeres to each other increases and they arranger themselves into a true eipthelium called *blastoderm*. This blastoderm remains two-ceil thick towards the animal pole of the egg and here, it forms a sort of root of the blastocoel. The sides and floor of the blastocoel are occupied by multilayered blastoderm of large yolky blastomeres. The resultant embryo having fluid filled blastocoel and blastoderm is the *blastula* of frog.

Fate Maps of Blastula of Frog

In the blastula, the blastomeres which have to form different germinal layers and different organs of the adult frog, have their representation at the external surface of the blastula. The fate of each type of blastomeres has been observed by artificial-vital staining methods of *Vogt* (1925) and prospective organ region maps or fate maps have been prepared for the blastulae of frog. According to the fate map studies, the whole surface of the blastula of frog can be divided into following three areas:

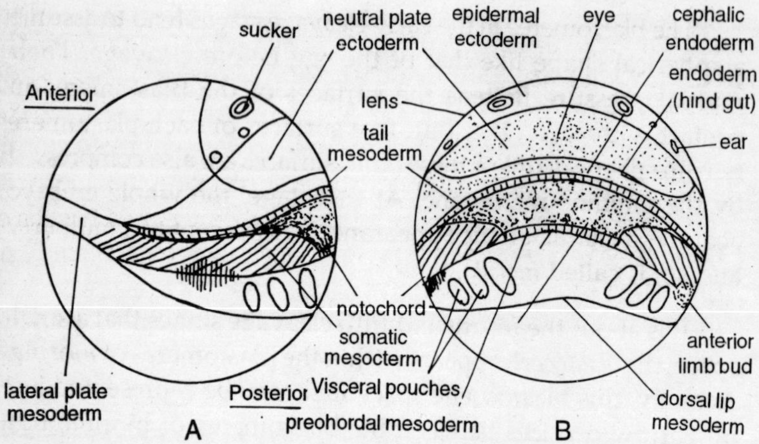

Fig. 4.6. Fate maps of blastula of frog. A - Viewed, from the lateral side. B - Viewed from the dorsal side.

Prospective Ectoderm Area

The entire pigmented area on and around the animal pole is the prospective *ectoderm*. The *neural ectoderm*, which develops into the central nervous system, occurs largely on the future dorsel side of the blastula, while the *epidermal ectoderm*, which develops into the skin epidermii of the embryo occupies the anteroventral side of the blastula. The material for the sense organs is also contained in these two areas. Inside the neural ectoderm area occurs a small sub-area which develops into the eye of the embryo. Likewise, inside the epidermal ectoderm area occurs the sub-areas for nose, sucker, ears and ectodermal part of mouth.

Gilchrist (1968) has called the zone of prospective ectoderm as *"zone of expansion,"* because, during gastrulation, this zone expands downward and then, converges toward the vegetal pole. At the end of gastralation, the ectodermal zone gives rise to the entire outer layer of gastrula.

Prospective Nnotochord and Mesoderm Area

The pigmented animal hemisphere (the prospective ectodermal area) is followed by a crescentic area of gray colour, the *marginal zone,* which goes all along the equator of the blastula, and has blastomeres for the formation of *notochord* and *mesoderm* of the frog embryo. The notochordal cells occupy the large area of dorsal side of the gray crescent. Beneath the notochordal area, and toward the vegetal pole, lie a narrow strip of cells which from the pre-chordal-connective tissues of the embryo and hence, this strip is called pre-chordal plate. At the upper side of notochordal area and toward the animal hemisphere, lies another narrow strip of cells which develops into the tail mesoderm of the frog embryo. The parts of the gray crescent on both sides of the notochordal area are taken up by the material for the segmental muscles, i.e., somites. The lateral and ventral parts of the gray crescent give rise to the mesodermal lining of the body cavity, the kidneys, etc., (ventro-lateral mesoderm).

The gray crescent or marginal zone has been named as the *"zone of involution"*, for its fate is to roll around and beneath the lips of the blastopore.

Prospective Endodermal Area

The entire non-pigmented area on and around the vegetal pole of frog blastula is the prospective *endoderm.* Just beneath the prechordal plate area of mid-dorsal part of gray crescent, occurs the endodermal cells for the anterior parts of the alimentary canal — the endodermal lining of the mouth, gill region and pharynx. Rest of the endodermal area includes the materials for mid-gut, hind-gut and other organs such as liver, pancreas, urinary bladder and certain endocrine glands.

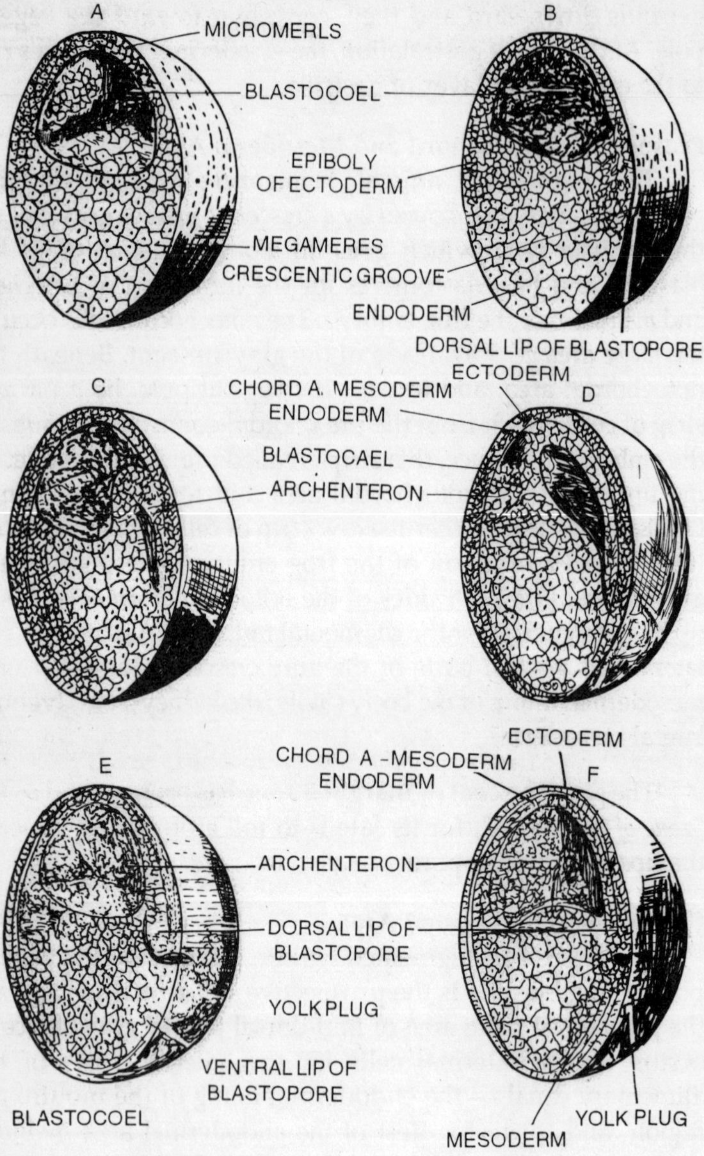

Fig. 4.7. Development of frog. Blastula and its transformation into gastrula; A — Blastula; B — Beginning of gastrulation; C to E — Elimination of blastcoel by the growing archenteron; F — Fully formed gastrula

Gilchirist (1968) has called the prospective endodermal zone, the *"zone of invagination"*, because, during gastrulation, it sinks and glides into the blastula of frog.

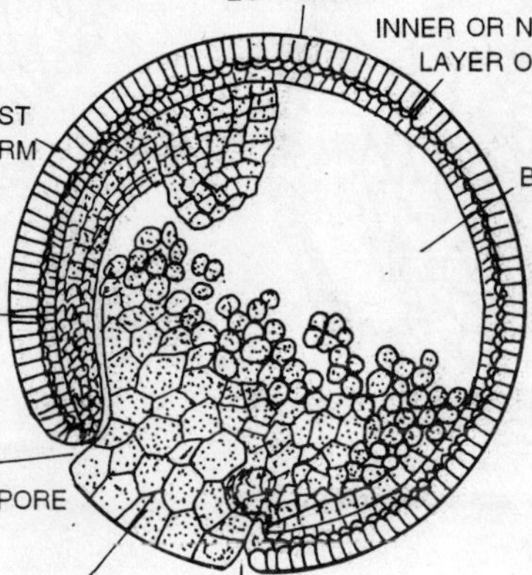

Fig. 4.8. Sagittal section through the embryo of frog.

GASTRULATION

Blastulation is followed by the complex process of gastrulation. In frog the gastrulation involves the following processes:

1. Epiboly

The pigmented cells of blastula have the tendency to overgrow the yolk cells. This process which is called epiboly begins slowly in the late blastula and accelerates with advancing gastrulation. The cells of the animal pole proliferate and grow downward on all sides except in the region where the dorsal lip of the blastopore has been formed.

2. Convergence

Vogt (1928) and Goerttler (1925) suggested that there is a streaming of the material of the dorsal and lateral surfaces of the early gastrula towards the blastopore. At the same time, their is a shifting of the lateral regions towards the mid-line.

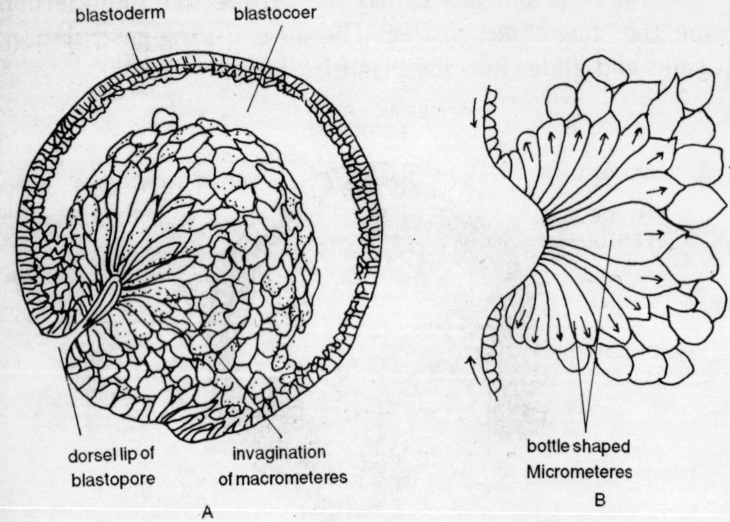

Fig. 4.9. Invagination of endodermal cells in an amphibian egg. A - Slightly schematized section through an advanced gastrula. B - The mechanics of gastrulation, showing active streaming of endodermal cells by bottling process

This combined type of movement is now generally described as *convergence*. Thus like amphioxus, here too the lips of blastopore do not actually constitute the sides the embryo. Holtfreter (1943) has tried to give physicochemical explanations in this movement. He suggested that an unfolding of denatured protein molecules is partly responsible. This unfolding, he thought causes a spreading of the superficial cells over a substrate with appropriate adsorption properties. Hence epiboly and convergence are due to this spreading tendency, which is apparently augmented by a lowering of surface tension in parts of the spreading cells.

3. Rotation

The epiboly continues untill the dorsal lip has passed over to an are a some what greater than 90°, and the area of white (blastopre) is reduced to small circle. This area will be situated rathe beyond the vegetal pole. Side by side, the entire developing gastrula has been rotating about a horizontal axis

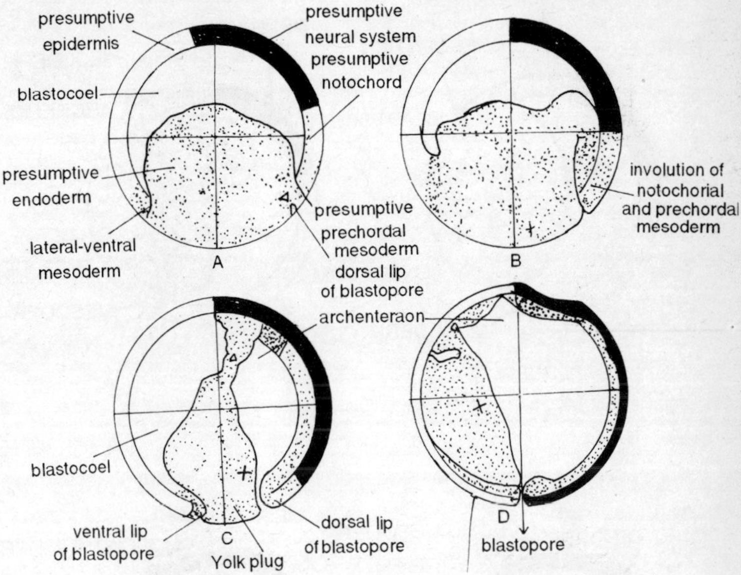

Fig. 4.10. Inward movements of the presumptive organ-forming areas of the blastula during the process of gastrulation in the frog egg. The triangle marks the endodermal cells which begin invagination and form the foregut. The cross marks the position of endodermal cells which were at the vegetal pole when gastrulation began and which were covered by epibolic growth of ectoderm.

lying at right angles to the original median plane of the egg. It means that the direction of rotation is such that the dorsal lip is in a sense carried backward in one direction as fast or faster than epiboly moves it forward in the other. The result will be that the blastopore, formed at approximately the vegetal pole, is posterior, and the morphologically dorsal and ventral lips are actually dorsal and ventral. This also makes is clear that the animal pole of the egg is to form the antero-ventral side of the future embryo while the region marked by the grey crescent is to form the dorsal side.

4. Invagination

Side by side to the above described external processes, a

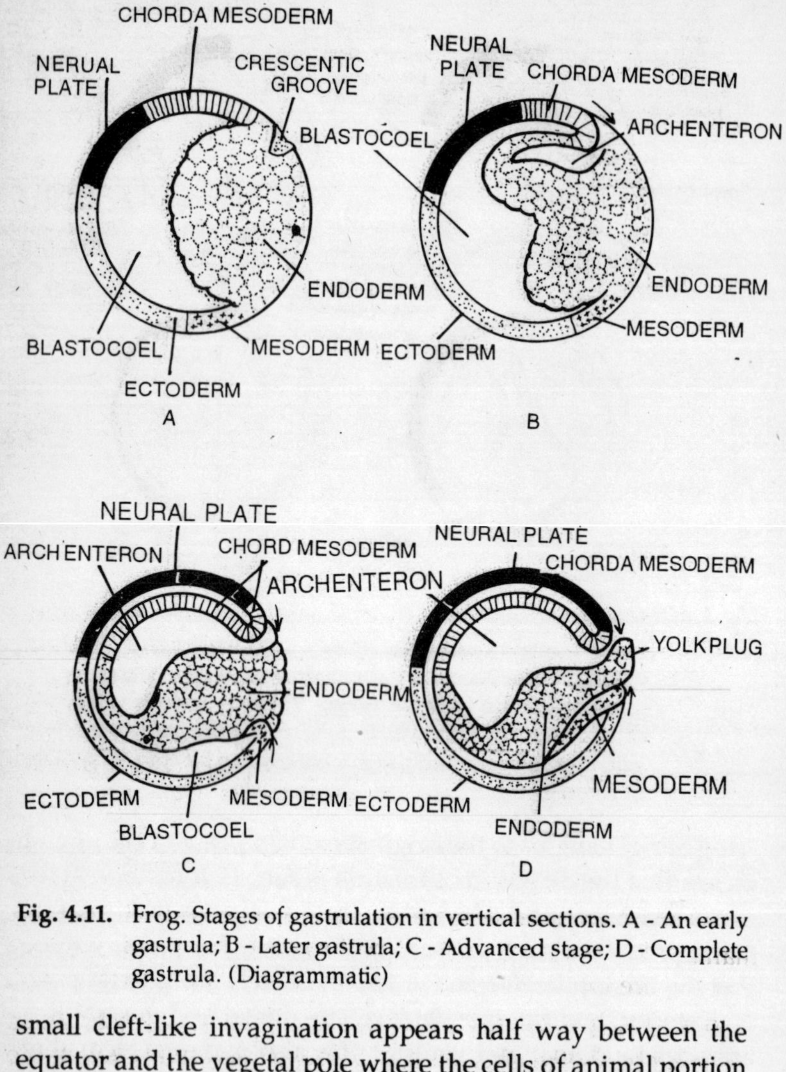

Fig. 4.11. Frog. Stages of gastrulation in vertical sections. A - An early gastrula; B - Later gastrula; C - Advanced stage; D - Complete gastrula. (Diagrammatic)

small cleft-like invagination appears half way between the equator and the vegetal pole where the cells of animal portion and vegetal portion merge. The upper or dorsal edge of the cleft forms the dorsal lip of the blastopore. This input swing begins on the dorsal side nearest the dorsal lip and spreads part way around the margins of the blastocoel in company with the external extension of the lateral lips. This modified invagination continues until the blastocoel cavity has been vertually eliminated, except for the narrow slit separating the outer layer of cells (epiblast) from the inner yolk cells (hypoblast). The new cavity formed is called the archenteron.

5. Involution

In addition to above processes a distinct *involution* also occurs at the blastoporal margin. This is most active at the median dorsal lip and progressively less so as one passes around either side untill at the ventral lip where there is always more at all.

In this process, cells located along the upper margin of the blastoporal lip move over the lip to the inside edge of the lip. There involuted cells are deposited on the inside of the embryo along the inner margin of the blastopore.

This means the roof of archenteron is composed of involuted cells *(presumptive choda-mesoderm)* and beyond this the outer layer (ectoderm). The floor and sides contain largely the endodermal cells, which have been derived from larger yolk cells, formerly located in the vegetal hemesphere of the blastula. In the later stages of the gastrula, the cells forming the floor of archenteron thin out considerably where they boarder blastocoel.

The crescent shaped dorsal lip of blastopore which appear at the beginning of the gastrulation continues to expand, first becoming semicircular, then horse-shoe shaped and finally forming into ring. This ring is the completed blastopore. Numerous yolk cells of vegetal hemisphere present is this region are crowded into the blastoporal opening where they form a mass referred to as *yolk plug*. Blastopore decreases in size while the archenteron is still in formation and when the later is completely formed, the yolk plug appears as only a minute oval or pear shaped speck on the gastrula.

6. Delamination

The word delamination denotes a mass separation of group of cells from other cell groups. It has been suggested that the separation of notochord, mesoderm and endoderm tissue from each other to form descrete cellular masses is done by the process of delamination, after these materials have moved to the inside during gastrulation.

In gastrula, thus the germ layers all establisted.

The various parts of the embryo develop from these germ players. the derivatives of the different layers are tabulated below:

Ectoderm	*Mesoderm*	*Endoderm*
Gives rise to the epidermal and its derivatives; to the nervous system; to the olfactory and auditory epithe lium; to the retina and lens of the eye; to other sensory organs; to the epithelial lining of the mouth and the anus; and to the pineal and pituitary body.	Gives rise to the connective tissue and muscles; to the skeleton except the notochord : to the blood vessels and lymphatics; to the peritoneum and the urinary and reproductive organs; dermis; parts of the eye except lens, cornea, and conjuctiva.	Gives rise to the epithelial lining to the gut and its diverticulae including the glands of oesophagus, stomach, intestine, bile duct, gall bladder, pencreatic duct, and the hepatic cells of the liver; respiratory tract; larynx, trachea and lungs; lining of urinary bladder; pancreas; thyroid and thymus.

Polarity and Rotation

From fertilization to early gastrulation, the frog's egg remain in the same position in regard to its polarity. After gastrulation its polarity changes. This is associated with the distribution of materials within the gastrula and can be accounted for by changes of the centre of gravity.

Closing of Blastopore

When the gastrula is formed, the blastopore appears as a small round opening filled with yolk plug. As it continues to diminish in size, it assumes a pear-shaped outline through the mutual approach of its lateral lips. Finally, these fuse completely to form a longitudinal groove, the *primitive streak,* ending dorsally and ventrally in a small aperture. Soon the lower aperture

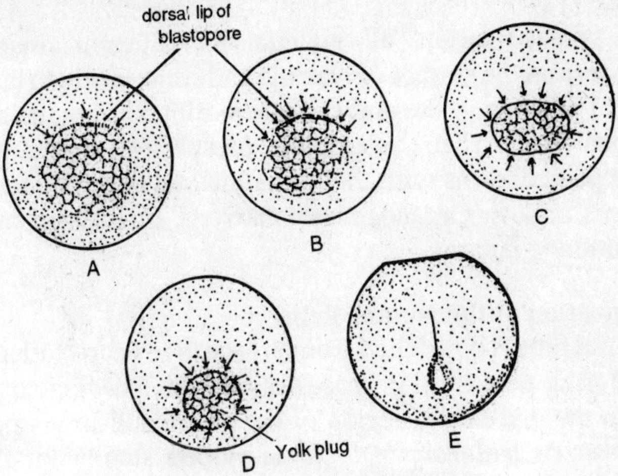

Fig. 4.12. Changes in the shape of the blastopore and its closure during gastrulation in frog.

closes, leaving only a dipression called the *anal pit*. The upper ones remain open for some long time and closes for later when the neural groove is laid down thus initiating the next figure' the *neurala*.

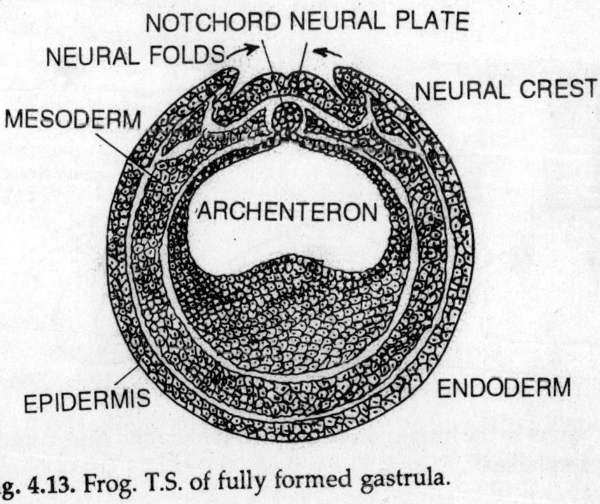

Fig. 4.13. Frog. T.S. of fully formed gastrula.

POST GASTRULATION OR ORGANOGENESIS

During pregastrulation, all material for different organs disappear from the surface of blastula and come inside to take their final position in the embryo where different organs are developed from their respective prospective areas. So *organogenesis* concerns with the formation and differentiation of various organs. Organogenesis converts an embryo into free swimming larva.

The Formation of the Neural Tube

Neural tube is developed from prospective neurectoderm which during pregastrular processes takes its position some where in the mid-dorsal region of the gastrula. Pear shaped *neural plate* or *medullary plate* is developed simply by the thickening of neurectodermal cells. The ectoderm on the sides of the ventral plates rises as a pair of *neural folds*. The edges of neural folds are developed by a process of differential growth and migration of cells called the *convergence*. The plate between the folds sinks down wards and form the *neural groove*. The elevation of neural folds increase dorsally and ultimately fuse over the neural groove to from the *neural* tube.

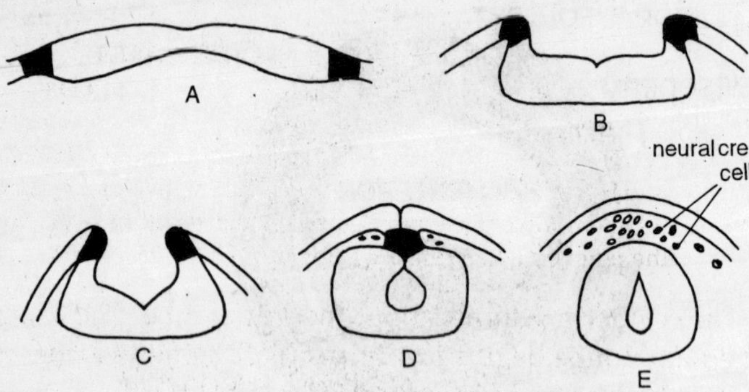

Fig. 4.14. Stages in the formation of the neural plate and neural tube in amphibians.

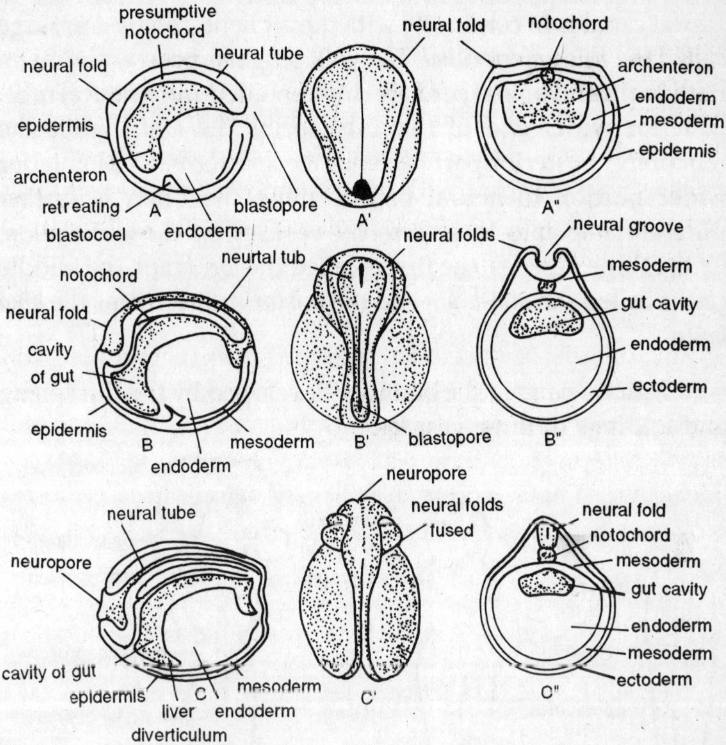

Fig. 4.15. Three stages of neurulation in a frog embryo. Figures (A′, B′, C′) show whole embryos in dorsal view; figures (A, B, C) show that right halves of embryos cut in the median plane. Figures (A″, B″, C″) show the, anterior halves of embryos cut transversely A, A′, A″, very early neurula; B, B′, B″, middle neurula; C, C′, C″, late neurula with closed neurula tube. C shows blastopore closed; the asterisk in C indicates the point at which the anal opening will break through.

The ectoderm reforms about the neural tube. The closure of the neural tube begins just in front of the mid-region and proceeds both anteriorly and posteriorly. At the anterior end the neural tube remains open as a *neuropore*. The neuropore persists only for some time and close ultimately. At the posterior end, the neural folds extend on the sides of the blastopore and meet over it. As a result of this enclosure of the blastopore, the neural canal is in continuity with the archenteron by a passage called the *neurenteric canal*. The yolk plug has been withdrawn. Both brain and spinal cord are differentiated from neural tube. Anterior wider end of this tube forms the *brain* while the remaining posterior part becomes the *spinal cord*. The anterior wider portion to neural tube forming the brain is further differentiated into three cerebral vesicles by three dialation. Of the three vesicles, the first is called the *fore brain*, the middle one is called the *mid-brain* and the posterior one forms the *hind brain*.

Various parts of the brain are developed by the thickening and foldings of three vesicles.

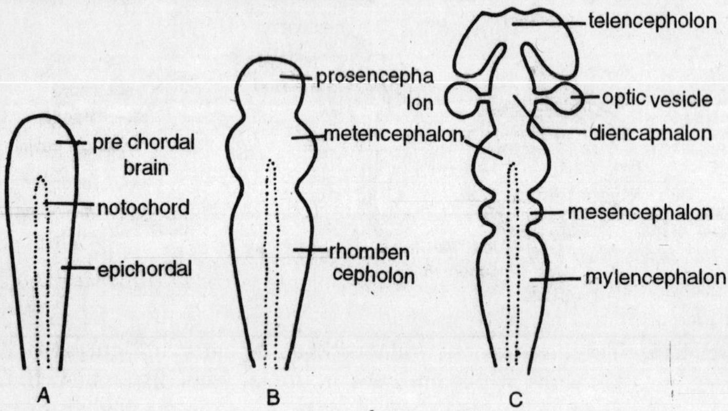

Fig. 4.16. Differentiation of various parts of brain.

During the differentiation of brain, there develops a liner band of cells forming *neural crest* on the side of the neural tube. This linear band of cells then divides into many small parts forming the *dorsal root ganglia* of cranial and spinal nerves. The arrangement of the ganglia is metameric and their positions correspond with those of the somites. Visceral arch skeleton and sympathetic ganglia are also developed by the proliferated cells of neural crest.

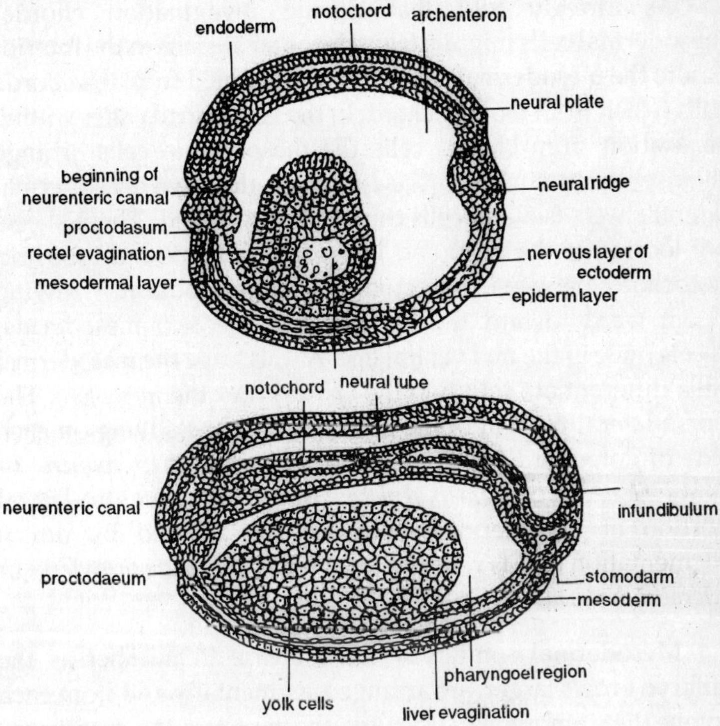

Fig. 4.17. Showing the notochord and neural tube.

Formation of Notochord

The *notochord* develops from the chorda-mesoderm, lying in the mid dorsal line of the gastrula. In the beginning, the cells forming the notochord are indistinguishable from the mesoderm but later on these cells separate from mesoderm by a narrow cleft, by a process of *delamination*. Then these separated chorda cells very quickly expand and arrange them selves in the form of a cylindrical rod and at this stage they become vacuolated. Even the intercellular material of the notochord becomes vacuolated. An elastic and fibrous sheath is developed around these cells. This sheath is called the *notochordal sheath*.

The Mesoderm and its Derivatives

As already told that during invagination chorda-mesodermal cells migrate from the outer surface to the interior where the mesodermal cells become separated from the chorda cells which from the notochord. In the late gastrula after getting separation from chorda cells the mesodermal cells arrange themselves in such a precise manner that two mesodermal lateral sheets (several cells thick) are developed. These sheets are located at the right and left sides of the neural tube and notochord between the ectoderm and endoderm. Growing down ward toward the ventral side, the two mesodermal sheets unite in the mid ventral line. At this stage the mesodermal cells differentiate into two the *epimere* and the *hypomere*. The dorso-lateral portion of the mesodermal sheets lying on each side of notochord forms the *mesodermal somites* or *epimeres* or *dorsal mesoderm* or *segmental plate*. The remaining ventro-lateral portion of mesodermal sheets, is not affected by dorsal segmentation and is referred to as the *lateral plate mesoderm* or *lateral mesoderm*.

Mesodermal somites which increase in number as the embryo grows larger, are arranged segmentally and from each somite three embryonic structures, the *myotomes*; the *dermatomes*; and the *sclerotomes* are developed in the following way:

After the formation of mesodermal somites from mesodermal sheets, each somite splits in two halves. That portion of this splitted somite lying against the body wall or ectoderm forms the dermatome while the remaining portion of mesodermal somite constitutes the myotome. At the same time a number of cells are proliferated from mesodermal somites that constitute the sclerotome from which the axial skeleton is developed. From dermatomes, the dermis of the skin is developed and from myotomes most of the body muscles are developed.

In the due course of time, the lateral mesoderm is separated into two layers, an out one of *somatic mesoderm* which lines the ectoderm of the body and an inner of *splanchnic mesoderm* which lines the gut. Both the layers are separated from each other except at their free ends, by a cavity which forms the *coelom* of the organism. When the two local lateral mesodermal plates unite in the *midventral line*, the lateral coelomic cavities also unite to form a single continuous coelom. This coelem is also called the perivisceral cavity or splanchnocoel. From this splanchnocoel, the *pericardial cavity* is developed. Inside the pericardial cavity *heart* develops. Kidney is developed from lateral mesodermal plate thus enclosing a part of body coelom. The only region where the coelom does not become continuous is the area dorsal to the gut or ventral to the notochord. Here at this region the splanchnic mesoderm of right and left side come together and form the double membranous suspensory mesentery for the gut; the *dorsal mesentery*.

Development of Heart

Heart is mesodermal in origin specially from splanchnic mesoderm. From splanchnic mesoderm and probably from endoderm even, a number of cells are proliferated. These cells arrange themselves to form a tubular structure. This tubular structure is the *endocardial tube*. The cells of the endocardial tube constitute the innermost layer of the heart. This tube is covered in the beginning from lateral sides and later on from all the sîdes by *epimyocardium*. From this epimyocardium both second and third or middle and outer most layers are *epicardium* and the *myocardium*. The myocardium thickens to form the muscular wall of the heart. In the early stages the heart is suspended in the pericardial cavity by a mesentery the *dorsal mesocardium*. The heart is located free in the coelomic cavity, formed above by the splanchnic and below and laterally by the somatic mesoderm. This space is the *pericardial* cavity and the enveloper of this cavity is the *pericardium*.

The heart thus developed, is in the form of a simple tube which in the latter stages of development becomes S shaped in the pericardial cavity.

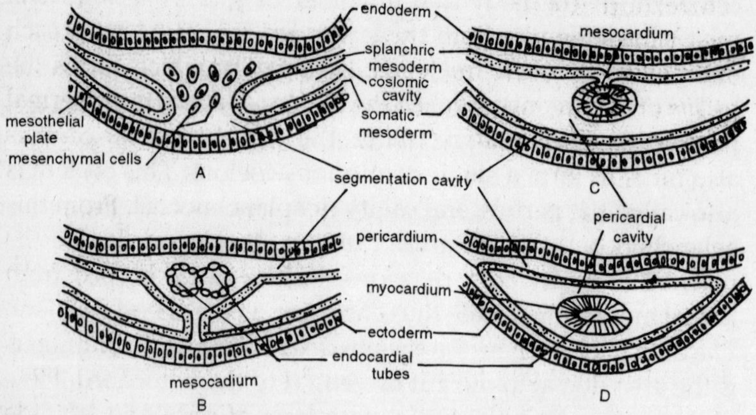

Fig. 4.18. Development of Heart.

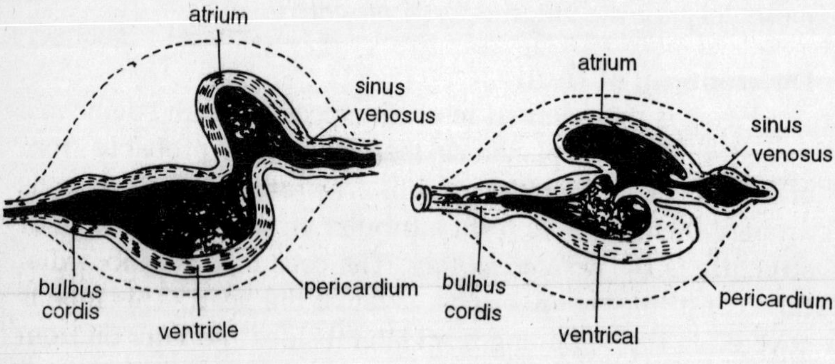

Fig. 4.19. Differentiation of various regions of heart.

The lumen of the heart is divided into chambers by the development of *valves* which are originated by the reduplication of the living membrane. Externally divisions of heart is due to the formation of some constrictions and valves resulting the formation of different regions of the heart such as *sinus venosus, atrium, ventricle,* and *conus arteriosus.* Out of these four, first two are the receiving chambers as they receive the blood from the body organs while the latter two are the distributing chambers and the blood is distributed to every part of the body through them.

Formation of Gut

During post gastrulation, the digestive tract or gut is developed from the archenteron of the pregastrula. During differentiation, the oesophagus and stomach are developed from the anterior part of the archenteron, while the posterior region differentiates into intestine. Pharynx oral opening, cloaca and anal opening are formed in the following fashion:

Two processes; invagination and evagination, controll the formation of pharynx mouth, rectum etc. At the time of hatching of larva an ectodermal pocket is developed by a process of invagination and at the same time another pocket or pouch is developed from endoderm lining the archenteron by a process of evagination. Then both ectodermal pouch and endodermal pouch meet to from *oral plate* which generally ruptures to from the *mouth. Pharynx* is developed by the evaginated portion of endoderm.

Anal region of hind gut is also developed by the same fashion. The endoderm ventral to the closed blastopore evaginates to fuse with the invaginating proctodael ectoderm to form the *anal plate.* This plate ruptures to form the *anus. Rectum* is developed by the evaginated portion of endoderm. The portion between rectum and anus forms *cloaca* which is lined by ectodermal cells.

Shortly behind the stomach, the yolky floor of the gut forms the *liver diverticulum* and its connection persists in the adult as the *bile duct*.

Development of Eye

Eye is developed from three different sources they are; the brain; the mesenchyme cells of the head; and the epidermis of the embryo. During the development *optic vesicles* are developed on either side of the brain simply by the out pushing of the ventro lateral part of the fore brain. The connection between optic vesicle and brain is the *optic stalk*. In the due course of time *lens* is developed by the thickening of epidermis

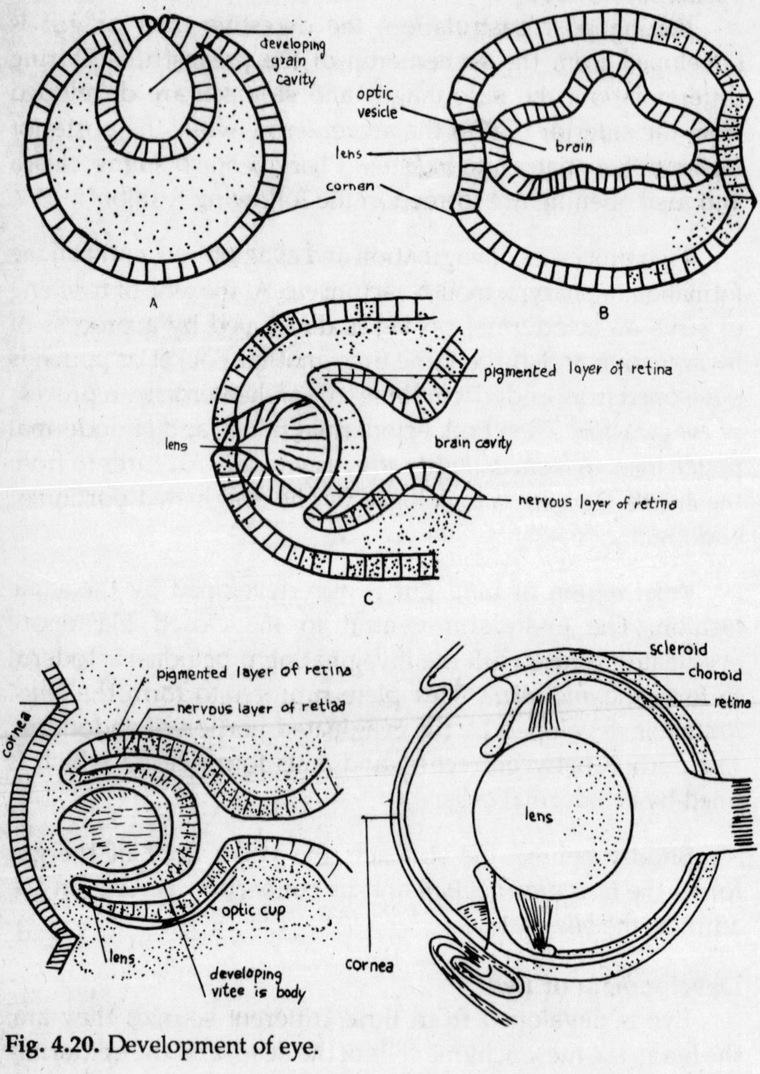

Fig. 4.20. Development of eye.

on the side of the embryo, opposite the optic vesicle. Actually this is done by invagination. Then a portion of the layer of optic vesicte invaginates to form two layered *optic cup* in which the lens comes to lie in. The portion of the layer of optic vesicle invaginates, constitutes the *nervous layer* of the retina while the portion of optic vesicle not invaginated constitutes the *pigmented layer* of retina. After some time both nervous layer and pigmented layer of retina come to lie in contact with each other. The retina and lens become enveloped by the mesenchyme cells of the head regions. From these mesenchyme cells *choroid, scleroid* and *cornea* are developed. A set of six eye muscles is originates from the mesodermal somites of head.

Further differentiation of pupil choroid fissure, optic nerve and optic chiasama etc., take place before metamorphosis.

Embryo of Frog

In vertebrates, the earliest stages in development may be modified by the amount of yolk that is present in an egg and also by the presence or absence of faetal membranes. The embryos of all vertebrates, particularly air-breathing, are surprisingly similar by the time of neurulation. This fact remains true for the embryos even if their habitate is water (amphibian), air (reptiles and birds), or the mother uterus (mammals). Moreover, the later neurula stage is of great importance in the evolution of the vertebrates. The time needed for the development of an embryo depends upon (*i*) the temperature, and (*ii*) the amount of oxygen present in the surrounding water.

The early is well-recognised by its larger size of structures. The head is well-formed and quite distinct while the tail has just started. By that time, the heart starts beating and a circulation of blood is maintained in the gills. The embryo is also provided with a conspicous ventral sucker to be used for attachment by the hatched embryo. With all these structures developed, the embryo comes out of the jelly which is dissolved by the enzyme secreted by the embryo itself. The stomodeum is formed and mouth starts developing. The embryo is capable of contracting its muscle when stimulated.

The young embryo is tubular and consists of following elements:

1. Outer epidermal covering.
2. Dorsal neural tube.
3. Notochord.
4. Endodermal canal or gut.
5. Mesodermal plate of right side.
6. Mesodermal plate of left side.

Besides these, there are also some loose cells inbetween these structures, which are called the mesenchyme cells. The mesenchyme, no doubt, is developed from mesoderm, yet it has some contribution from ectoderm also. These cells form the supporting tissues, or skeletal system, musculature, blood vessels and the blood. At this stage, the embryo of frog resembles the organisation of typical vertebrate embryo.

The Elongation of the Embryo

The embryo, till the gastrulation is completed, is almost spherical or oval but after this, it undergoes elongation which is typically in anterio-posterior direction. With the so-called elongation of body beyond the anal aperture, tail formed and in this tail region the notochord, nerve cord, somatic mesoderm mesenchyme (excluding endoderm) are given in. This post-anal is oftenly called as the *tail bud*. But this term is not proper as this position is formed by the rearrangement of tissues present in and near the hinder end of gastrula. There is no proliferation of indifferent cells around the margin of almost closed blastopore. This significance of formation of tail is that their stretching of posterior past of neural plate and neural folds along the longitudinal direction. And this is recently proved that somatic mesoderm of the tail develops from the posterior part of neural plate which may be called as prospective tail somites. Then thus-formed somites of tail are enclosed within by the hinder part of neural folds and thed elongates to a great extent. This hinder and develops a dorsal skin fold or fin and a ventral fin and thus the tail is fully formed.

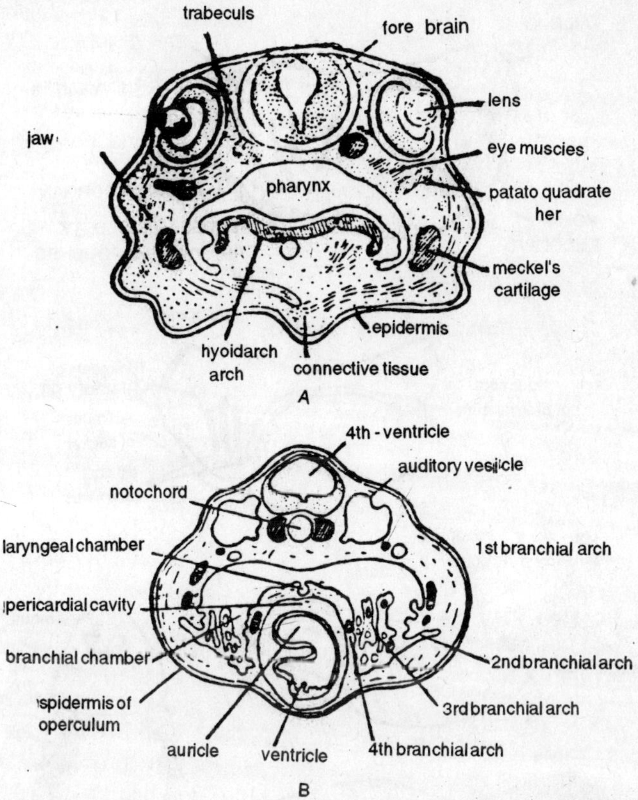

Fig. 4.21. Transverse sections passing through different regiong.

External Gills

Just after the embryo is hatched, the skin present on the dorsal part of the first and second and later on that of third branchial arch (i.e. visceral arches third, fourth and fifth) grows out to form three pairs of feathery *external gills*. These are the first formed gills and function as the respiratory organs.

Mucous Glands

When the external gills are developing, at the same time, on the other hand, ectoderm developes a deep pit lined with glandular cells. These pits are present on the ventral end of each mandibular arch which is the between the

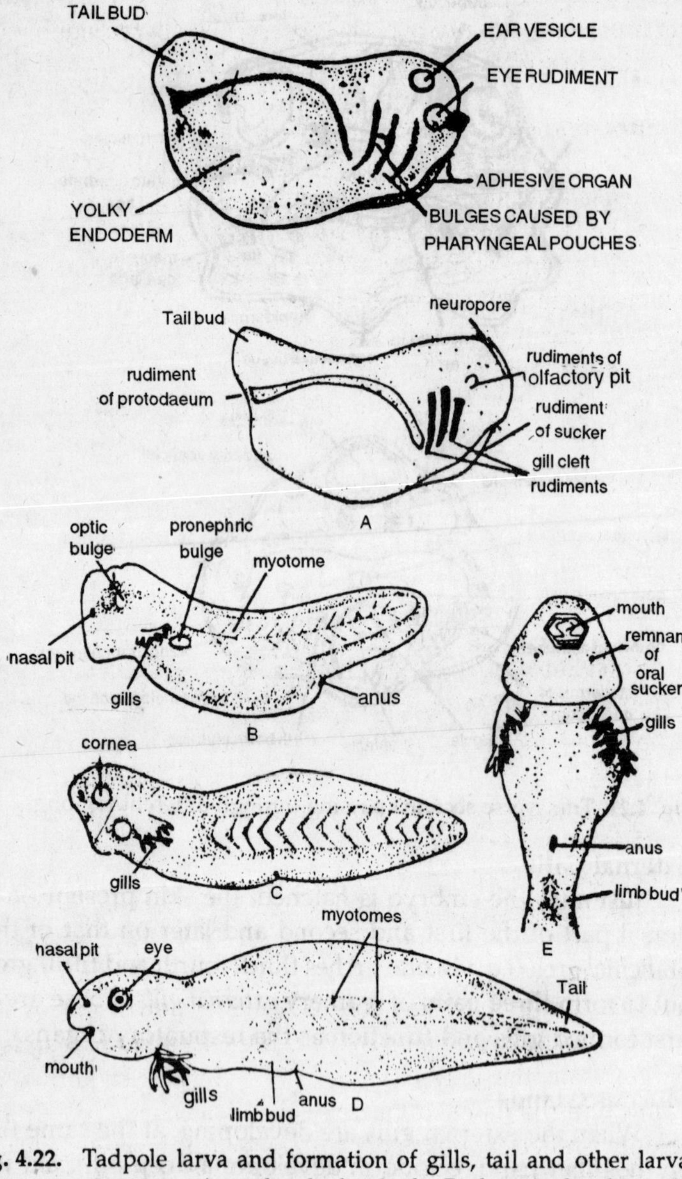

Fig. 4.22. Tadpole larva and formation of gills, tail and other larval structures. A - unhatched tadpole; B - larva at hatching; C - 8 mm long tadpole; D - 11 mm long tadpole; E - ventral view of 11 mm long larva.

hyomandibular pouch and the stomodeal invagination. These pits are called the *mucous glands* which help the embryo in attaching to some substratum, after hatching.

A. Formation of Tadpole Larva

About four days after fertilization, the embryo is said to be at the *tail bud* stage of development when it is about 3 mm. long. The tail bud larva undergoes a series of changes in its external and internal structure within the egg membranes to become the tadpole larva. The larva elongates in an antero-posterior direction and shows distinct differentiation into *head* and *trunk*. At the hind end appears a small outgrowth or *tail bud* indicating the beginning of the tail. Small swellings on each lateral side of the head indicate the position of the internally developing *eyes*. In front of each bulge is present a small *nasal pit*, and behind the nasal pit of each side lies the rudiment of the *ear*. A 'V'-shaped glandular pit, the *adhesive organ* or *cement gland* develops just below the prospective *stomodaeum* on the ventral surface of the head.

In the anterior part of the trunk, three *pharyngeal pouches* are formed as lateral outpushings of the fore-gut. These are the rudiments of first three gill clefts. The trunk bulges ventrally and laterally due to the presence of a large amount of yolk beneath and at the sides of the developing gut. A small depression, the *proctodaeum* lies between the tail and the trunk on the ventral side. It will form the *cloaca* and the cloacal aperture. The larva also has a well-differentiated brain.

B. Hatching

When the embryo is about 12 days old, it dissolves the egg membranes by *hatching enzymes* produced in the skin and emerges from the jelly envelope. This process is known as *hatching* and the hatched larva is called a *young tadpole*. It is a blackish, fish-like creature, about 7 mm long. The newly hatched larva swims for a short while, but soon fixes itself to some floating vegetation or moss-covered walls of the ponds by means of a sticky secretion of mucous glands present in the adhesive organ. It does not take any food due to the absence of mouth and is nourished from the yolk present in the ventral bulge of the trunk.

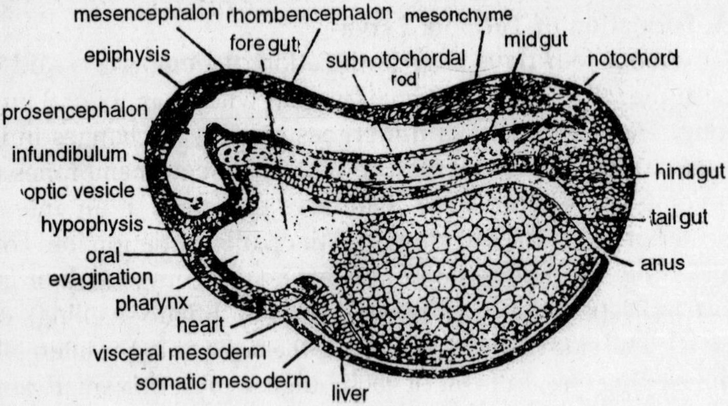

Fig. 4.23. Median longitudinal section of an early tadpole (about 3 mm long).

During the larval life of tadpole, following developments occur:

(1) The *mouth* forms.

(2) A pair of *horny jaws,* bearing *papillae* or *horny teeth,* develop around the mouth. The larva now begins to feed upon algae and other vegetable matter by the rasping action of the teeth.

(3) Three pairs of finger-like *external gills* project from the sides of head in the branchial region.

(4) The external gills gradually disappear under an opercular fold of the skin. Meanwhile, *true* or *internal gills* have been gradually developing by perforation of the pharyngeal gill pouches. The *opercular chamber* opens to the outside by an aperture called *spiracle* on the left side. The space between the body wall and operculum is called branchial or gill chamber.

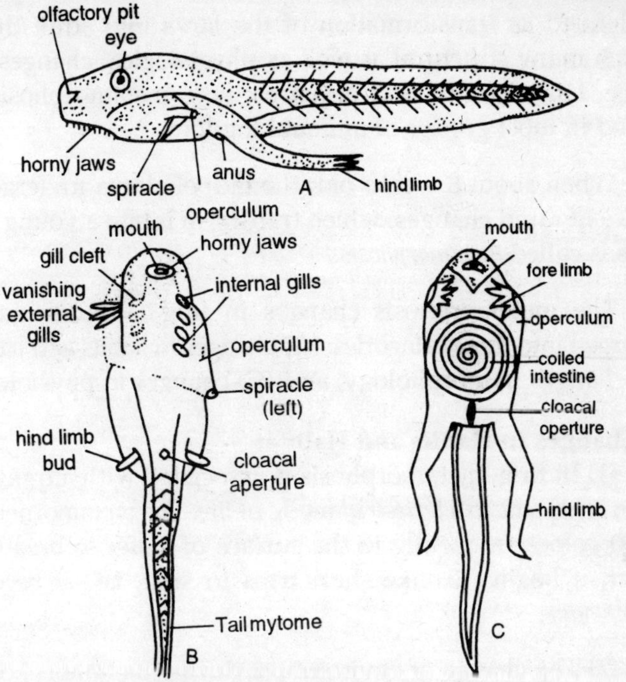

Fig. 4.24. A - Fully formed tadpole; B - Chance in respiratory organs; C - showing the large coiled intestine.

(5) The *tail* elongates and becomes a powerful organ of locomotion. It acquires a dorsal and a ventral *tail fin*.

(6) Adhesive organ begins to degenerate.

(7) Lens and cornea are formed and the *eyes* become functional.

(8) *Hind limb buds* appear, one on each side of the cloaca.

The embryo is now a free-swimming, actively freeding *tadpole larva*, which has no resemblance with the fish-like tailed semi-terrestrial adult frogs.

Metamorphosis

The term *metamorphosis* is derived from a Greek word *metamorphoun*, which means "to transform." In embryology it is defined as transformation of the larva into adult during which many structural as well as physiological changes take place. Larval forms and accompanying metamorphosis are found in most groups of animal kingdom.

When about 8 weeks old, the tadpole larva undergoes a series of rapid changes, which transform it into a young frog. This is called *metamorphosis*.

The metamorphosis changes in frog tadpoles may be grouped into three categories: (A) changes in habits and habitat, (B) changes in morphology, and (C) changes in physiology.

A.Changes in Habits and Habitat

(1) In frog, metamorphosis is associated with a transition from an *aquatic* to a *terrestrial* mode of life. The metamorphosing larva comes frequently to the surface of water to breath air. Later, it begins to take short trips to land, i.e., it becomes *amphibious*.

(2) The change of environment during metamorphosis is associated with a change in feeding. The tadpoles are *harbivorous* feeding upon algae and other vegetable matter — which they scrape off with the help of the horney teeth surrounding their mouths. Adult frogs, on the other hand, are *carnivorous* — living on insects, worms etc. Sometimes, they also devour larger prey, such as smaller frogs and even little birds and rodents which they overpower and swallow.

B. Changes in Morphology

(i) Regressive or destructive changes : These include the reduction or complete disappearance of those organs or structures which are necessary during larval life, but not essential in the adults. The important changes of this nature are as follows:

(1)The long *tail* of the tadpole along with the fin folds is absorbed again during metamorphosis and disappears altogether at the end of the metamorphosis.

The reduction of the tail is affected by *autolysis* of the component tissues with active participation of amoeboid *macrophages*. These are rich in lysosomes that contain a proteolytic enzyme *cathepsin*. The macrophages phagocytose the debris of the disintegrating cells. During metamorphosis, however, the tadpole is nourished by the substance of its tail.

(2) The *gills* are resorbed, the *gill clefts* are closed and the *branchial cavities* disappear. The resorption of gills also takes place by autolysis.

(3) The *horny teeth* of the perioral disc as well as the *horny lining* of the jaws are shed.

(4) The *lateral line sense organs* present in the skin of tadpoles, disappear during metamorphosis.

(5) The *cloacal tube* becomes shortened and reduced.

(6) Some *blood vessels*, including parts of the aortic arches, are reduced.

(ii) Progressive or constructive changes : These include the development of some organs which become functional only during and other metamorphosis.

(1) There is a progressive development of the *limbs* which increase in size and differentiation. The *forelimbs*, which in the tadpole develop under cover of opercular membrane, break through to the exterior. At the same time, there is a great increase in the length and strength of the *hindlimbs*. Joints develop in them and the toes appear.

(2) The *middle ear* develops in connection with the first pharyngeal pouch. The tympanic membrane develops. It is supported by the circular tympanic cartilage and enables the frog to receive air-borne vibrations.

(3) The *eyes* bulge up on the dorsal surface of the head and develop nictitating membrane.

(4) There is development of the *tongue* and the *vomerine teeth.*

(iii) Remodelling of some structures : Some organs which function both in the larva and the adult, change their differentiation during metamorphosis so as to meet the requirements of the adult mode of life.

(1) The *skin* of the tadpole is covered with a double-layered epidermis. The number of layers of cells in the *epidermis* increases during metamorphosis. Surface layers become cornified. Multicellular *mucous* and *serous glands* develop in the skin. The *pigmentation* of the skin is changed; new patterns and colours appear.

(2) There is a widening of the *mouth* gap due to rotation of the quadrate cartilage backwards and the *true jaws* become functional.

(3) The *tongue* grows much larger and more muscular.

(4) The *stomach* and the liver enlarge.

(5) The *eyes* grow more prominent.

(6) In tadpoles, the intestine is very long and would up into a spiral like a watch spring. The intensive becomes greatly elongated due to a *harbivorous* habit because the vegetable diet contains less nourishment than the animal diet. Hence, a longer gut is required to extract most of the nourishment. During metamorphosis, the intestine is greatly shortened and most of the coils which it forms in the tadpole become straightened out. This change is apparently correlated with the *carnivorous* habits assumed by the adult.

(7) Two chambered *heart* of frog is transformed into a three-chambered one.

(8) *Lungs* do not undergo major changes. They develop very gradually and become fully functional.

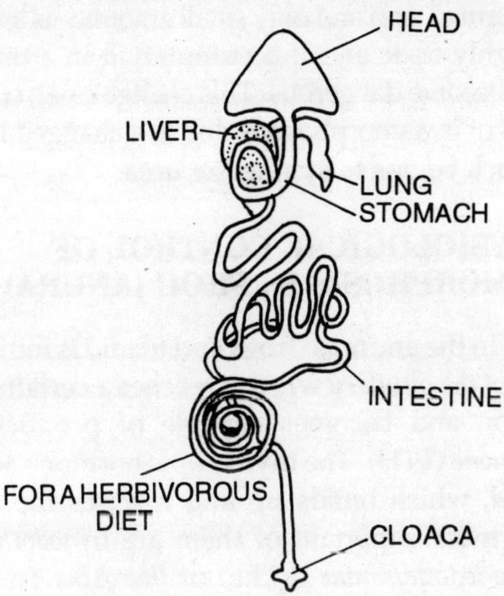

Fig. 4.25. Lengthy gut of a tadpole.

(9) The cartilaginous *skeleton* of tadpole undergoes ossification.

(10) The larval *pronephric kidney* is transformed into a *mesonephros*.

(11) The *blood vascular system* is modified with the disappearance of some aortic arches and increase of blood supply to the lungs and the skin.

C. Changes in Physiology

(1) The endocrine function of the pancreas starts at metamorphosis as it begins to secrete *insulin* and *glucagon* hormones. This is related to the increased role of the liver in the turnover of glycogen.

(2) A marked change takes place in the excretory function. In the tadpole, the end product of nitrogen metabolism is ammonia which is easily disposed off by diffusion in water. After metamorphosis, however, the frogs excrete most of their

nitrogen in the form of urea and only small amounts as ammonia. The latter is highly toxic and if accumulation in a terrestrial animal, it could become dangerous. This change-over, occurring in the late stages of metamorphosis is due to a changed function of the liver, which begins to synthesize urea.

PHYSIOLOGICAL CONTROL OF METAMORPHOSIS IN FROG (ANURA)

Metamorphosis in the anurans (frogs and toads) is initiated by the *anterior lobe of the pituitary* when it reaches a certain degree of differentiation and becomes capable of producing the *thyrotrophic hormone* (TTH). The thyrotropic hormone activates the *thyroid gland*, which builds up and releases the thyroid hormones. The most important of them are *tri-iodothyronine* (I_3Thn) and *tetra-iodothyronine* (I_4Thn) or *thyroxine*. In both of these compounds two residues of the amino acid *tyrosine* are joined together, and to these are attached 3 (in I_3Thn) or 4 (in I_4Thn) atoms of iodine. The structural formulae of these hormones are as follows:

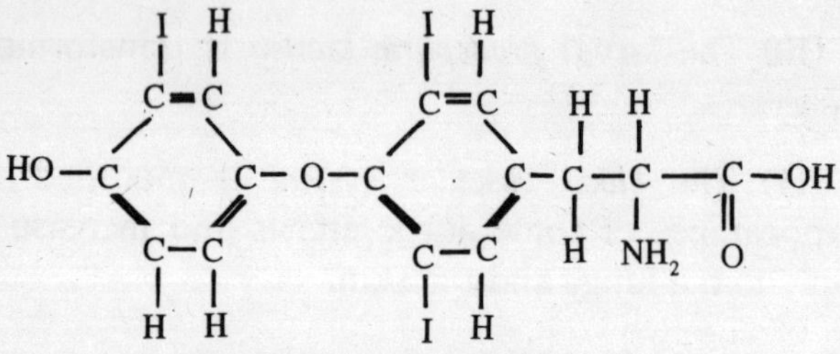

Tri-iodothyronine (I_3Thn)

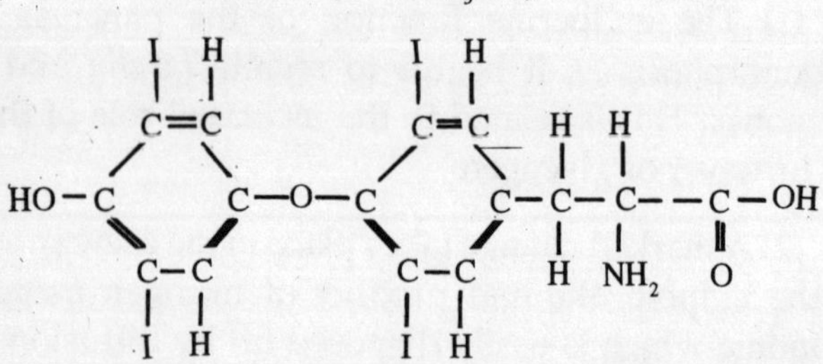

Thyroxine

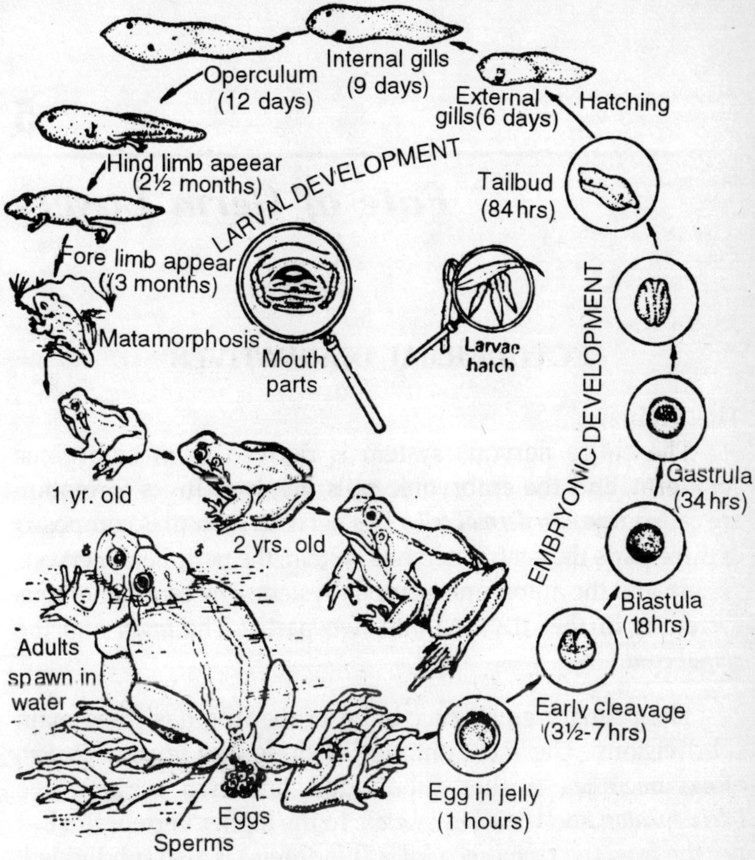

Fig. 4.26. Frog. Life cycle.

Of the two compounds, thyroxine is produced in much greater quantities and is thus the main action substance released by the thyroid gland, although the activity of tri-iodothyronine is three to five times as high as that of thyroxine.

The thyroid hormone is carried to the various body organs through the blood stream and affects the tissues directly. While it causes the degeneration and necrosis of some cells, it slimulates the growth and differentiation of others, thus causing the metamorphosis.

5

Fate of Germ Layers

ECTODERMAL DERIVATIVES

The Brain

The entire nervous system is derived from embryonic ectoderm, and the embryonic cells involved in its formation are called *neurecto dermal cells.* The nervous system is composed of three parts the central nervous system, the peripheral nervous system and the autonomic nervous system. The central nervous system is further divided into two parts : The *brain* and the *spinal cord.*

The primary embryonic brain of the frog has three main subdivisions. The most anterior of these, the *prosencephalon,* alone becomes further subdivided into two regions, the *telencephalon* and the *diencephalon.* In the higher vertebrates but not the frog, the *rhombencephalon* (hindbrain) is also subdivided. The adult frog brain then has four major divisions, arbitrarily determined, for they merge into one another structurally and functionally. Each of these divisions has a specific set of characteristics and derivatives, which will be described in sequence from the most anterior to and including the spinal cord.

The Prosencephalon

It is the anterior most hollow part of the neural tube that forms the prosencephalan or fore brain. The prosencephalon is so strongly curved down words in the early embryo that it is almost at right angles to the remainder of the neural tube.

Shortly after hatching, the prosencephalon differentiates more clearly into parts — the telencephalon and the diencephalon. The *telencephalon* with its original cavity, the *telocoel*. One may draw a line in a sagittal plane from a point just anterior to the epiphysis and extending through the brain cavity to the posterior wall of the thickened *torus transversus*. This imaginary line separates the anterior telencephalon from the more posterior diencephalon of the forebrain. The telencephalon includes the fused anterior neural folds which comprise the *lamina terminalis*. The lamina will form a partition which will separate the paired cerebral hemispheres by a longitudinal groove.

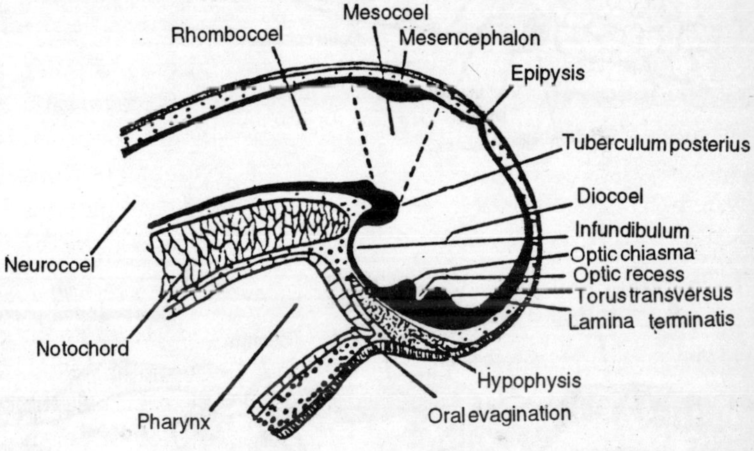

Fig. 5.1. Early organogeny of the frog tadpole, showing the primary brain vesicles in sagittal section.

The telencephalon is the embryonic *cerebrum* and differentiates as a thin walled vesicle. Its cavity expands laterally to give rise to the right and left lateral *ventricles* and the surrounding thick-walled *cerebral hemispheres* or *lateral telencephalic vesicles* at about the 12 mm stage. These ventricles are laterally compressed. In the frog the cerebral hemispheres are first differentiated at the 7 mm. stage but never become very large. The two telencephalic vesicles are partially constricted off from each other but remain connected by way of the tubular *foremen of Monro*, which opens into the common (intermediate)

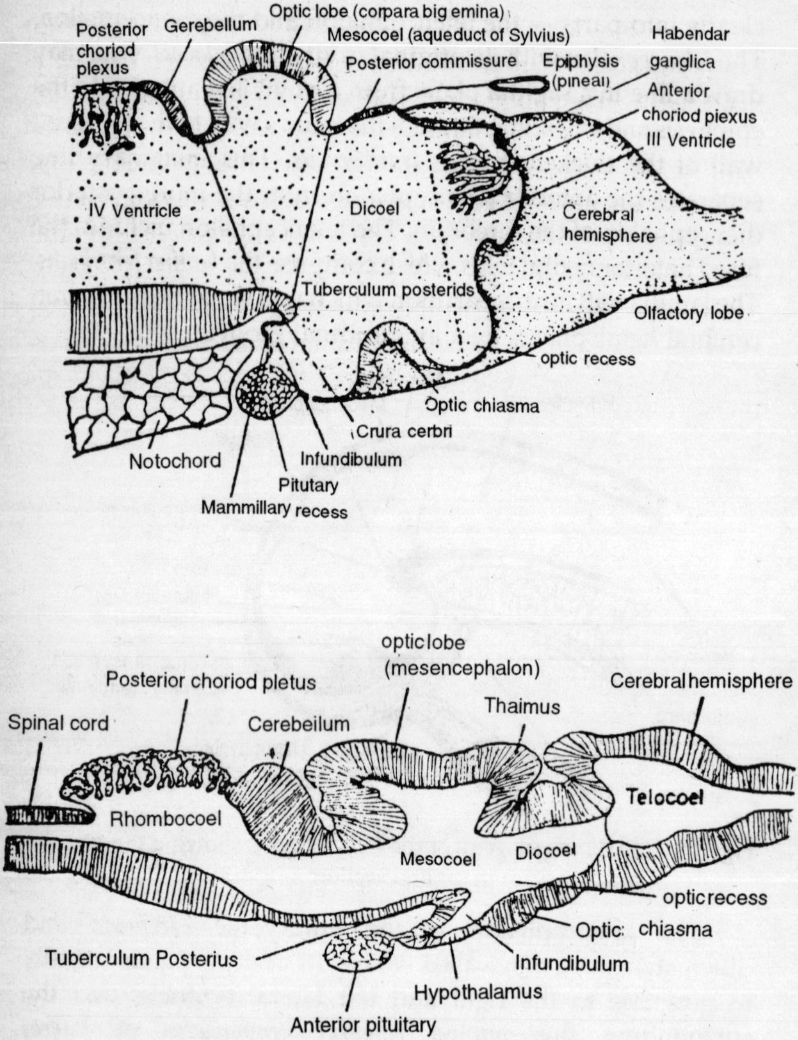

Fig. 5.2. Late development of the brain of the frog tadpole. (*Top*) Premetamorphic stage. (*Bottom*) Adult brain, schematized, reduced in size.

third ventricle, known as the *ventriculus impar*. The third ventricle overlaps and connects the telocoel and the diocoel. The nervous tissue of the cerebral hemispheres is connected transversely by the ventral *anterior commissure,* which is the original torus transversus, and the *anterior pallial commissure* found posteromesially. The cerebral hemispheres becomes greatly thickened with consequent enlargement and there is a parallel reduction in the size of the contained ventricles. Antero-ventrally each grows out toward the anteriorly placed olfactory placodes to establish connections known as the *olfactory lobes* and *nerves.* The olfactory lobes arise as a pair but subsequently become fused mesially. These are rather prominent brain structures in the adult frog. They seem to originate together and become separated only at the point of origin of the *olfactory nerves.* However, each originates from the telencephalon on its side of the brain.

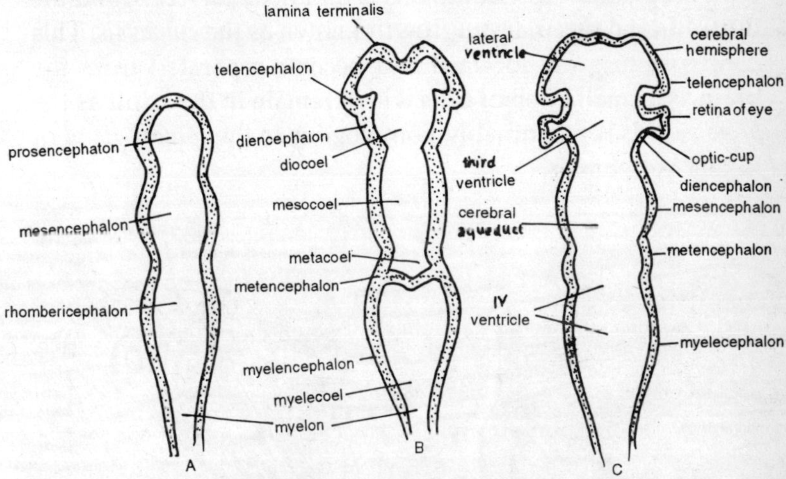

Fig. 5.3. Development and formation of basic parts of frog's brain.

Anterior to the epiphysis, the roof of diencephaless becomes convoluted and is supplied with a profuse blood capillaries forming the so-called the *anterior choroid plexus* which develops as thin and highly vascular invagination from the roof just at the junction of the telencephalon and the diencephalon. It extends into the diocoel and into the two telocoels, and is therefore tri-radiate.

The Diencephalon

The posterior portion of the prosencephalon, called the Diencephalon, retains its their roof. The diocoel, which will form the *third brain ventricle* is divide between the telencephalon and the diencephalon, but the enlarged portion found surrounded by the diencephalon is identified as the *diocoel.* This cavity extends antero-dorsally beneath the epiphysis, antero-ventrally into the optic recess, postero-ventrally into the infundibulum, and laterally into the relatively large optic *vesicles.*

The structural derivatives of this diencephalon include the *posterior commissure,* just anterior to the dorsal limit of the mesencephalon. Anterior to this is the *epiphyseal recess,* and the dorso-mesial saccular outgrowth known as the *epiphysis.* This continues to grow forward and become separated from the brain as a small knob of cells which remain in the adult as the *brow spot.* It is presumably homologous to the *pineal gland* of higher vertebrates.

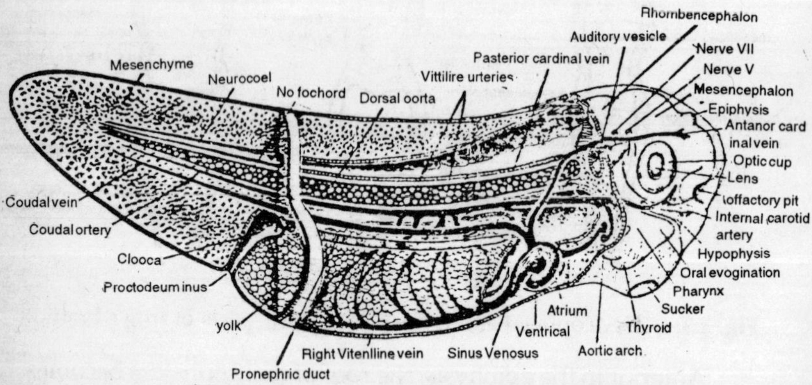

Fig. 5.4. Reconstruction of the 7 mm. frog larva showing the major organ systems from the right side.

Anterior to the epiphysis, in the roof of the diencephalon and between it and the anterior choroid plexus, are the *habenular ganglion* and commissure. In front of this there later develops a dorsal outgrowth know as the *paraphysis*. In the floor just posterior to the *optic recess* is the very thick *optic chiasma* which will contain crossed fibers from the two optic stalks. Posterior to this chiasma is a thin-walled pocket or trough projecting beneath the anterior tip end of the notochord, known as the *infundibulum*. The cells of the infundibulum will combine with the approximated and pigmented cells of the ingrown *hypophysis* to form the *pituitary gland* of the adult. The infundibulum cells give rise to the posterior part of the pituitary gland and retain a hollow infundibular stalk connection with the brain. The hypophysis becomes the anterior part of the pituitary gland. During metamorphosis the individual lobes of the pituitary gland differ, both in gross morphology and in finer structure. The lateral and intermediate lobes show poor vascularization, a simple cellular structure, uniform staining reaction, and a rather static structural development. The anterior lobe becomes highly vascular, consists of several types of cells, and exhibits increasing complexity with further development. The acidophils of the anterior pituitary gland become highly differentiated and the basophils become poorly differentiated when the thyroid gland is either poorly developed or inactive. Conversely, thyroid activity is correlated with hyperactivity of the basophils of the anterior pituitary gland. This gland is formed from ectoderm, but some is epithelial and some is brain ectoderm. Between the infundibulum and the tuberculum posterius is a secondary and posteriorly directed pocket known as the *mammillary recess*.

The *optic vesicles* begin to develop very early as lateral outgrowths from a slightly ventral level of the diocoel. The expansion of the diocoel provides a temporary and slight thinning of the walls of the optic vesicles. However, as these vesicles make contact with the lateral head ectoderm, that portion of the vesicle in contact begins to thicken and then invaginate to form a 2-layered optic cup. The most lateral and invaginated portion of the cup will become the *retina*. The mesial layer will become the *pigmented layer* of the eye, and the connecting and somewhat constricted tube the *optic stalk*. The nervous elements of this optic stalk will join in the *optic chiasma* which contains the *optic nerve* fiber tracts from the two sides. The stalk will develop around an inverted groove (the choroid fissure) which will contain, within the groove, accessory nerves and blood vessels which feed the retina.

The Mesencephalon (Midbrain)

The mesencephalon is the most conservating part of the vertebrate brain, undergoing the least amount of change. It separated from the diencephalon by the mesodien cephalic groove.

This portion of the brain function largely as a pathway of nerve tracts between the anterior prosencephalon and the posterior rhombencephalon. These tracts are found principally within the paired ventro-lateral thickenings of the walls and floor on either side of the tuberculum posterius. They are known as the *crura cerebri.*

The original dorsal thickening becomes subdivided by a median fissure into paired dorso-lateral thickenings, at about the 10 mm. stage. These are known as the *optic lobes* or *corpora bigemina.* They do not reach their full development until the time of metamorphosis. Anterior to these lobes is the *posterior commissure.* Two pairs of efferent nerves take their origin in the mesencephalon. The third cranial nerve or *occulomotor,* grows out from its neutral side, and the forth, or *trochlear nerve* originates dorso laterally, near the function of the mesencephalon with the rhambencephalon. The occulomotor supplies four eye muscles namely : *the superior rectus, inferior rectus, internal ractus* and *inferior oblique.* The trochlear supplies the superior oblique and the original cavity of the mid-brain (mesocoel) connects the *rhombocoel (fourth ventricle)* with the third ventricle which becomes narrow and is known as the *aqueduct of Sylvius* also the *iter etertio ad quartum ventriculum.*

The Rhombencephalon (Hindbrain)

This portion of the brain is clearly marked off from the

mesencephalon by a transverse constriction in the roof of the brain, at the posterior limit of the dorsal thickening. It is called the *mesoaneta çephalic groove.* It is not clearly divided farther. There appears a slight transverse thickening in the roof of the rhombencephalon which corresponds to the metencephalon of higher forms and develops into the *cerebellum* that remains very small in frog. Posterior to this the roof becomes broad, thin, and vascular, and folds into the *rhombocoel* (fourth ventricle) as the *posterior choroid plexus.* The ventral and ventro-lateral walls of the rhombencephalon are known as the *medulla oblongata* from which arise the cranial nerves V to X inclusive. The walls becomes thickened by fibers which form numerous pathways from the brain and cord.

The *rhombocoel* or cavity of the hindbrain is known as the *fourth ventricle* which communicates posteriorly with the *central canal* of the spinal cord also called *neurocoel* and anteriorly with the *aqueduct of Sylvius* of the mesencephalon.

The Spinal Cord

The myelencephalon is continous with the spinal cord which extends through the entire body to the tip of the tial. In the young tadpole it is oval in outline in transverse section having thick lateral walls and a thin roof and floor.

The neural or central canal (*neurocoel*) from the beginning of its development is laterally compressed by the thick lateral walls of the spinal cord derived from the original neural folds. The lining cells, coming from thedorsal epithelial ectoderm, retain both their pigment and their cilia, and are non-nervous in function. These cells, which continue to line the central canal of the adult, are known as the non-nervous *ependymal cells*. The thick lateral walls of the spinal cord are made up of the rapidly multiplying germinal *neuroblasts* (primitive or embryonic precursors of the neurons) and the supporting small and stellate cells known as the *glia (neuroglia)* cells. These cells have the function normally ascribed to connective tissue, namely support for the neuroblasts, but they are of ectodermal origin.

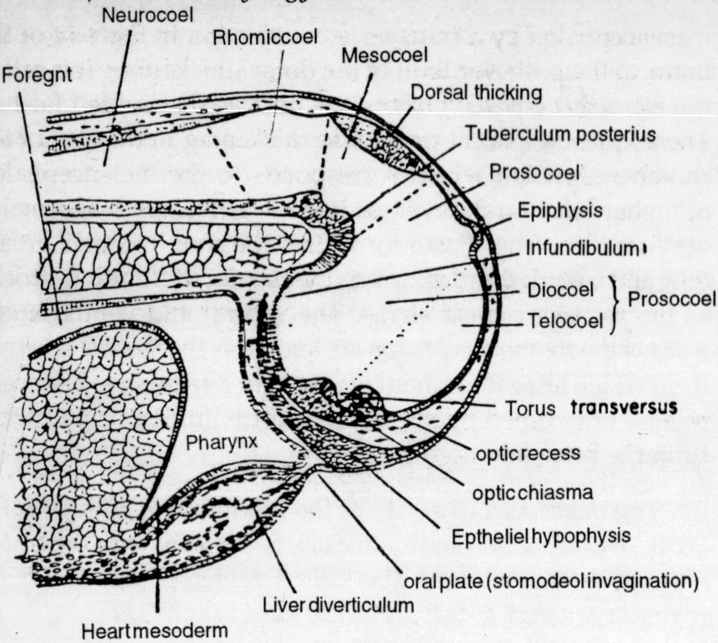

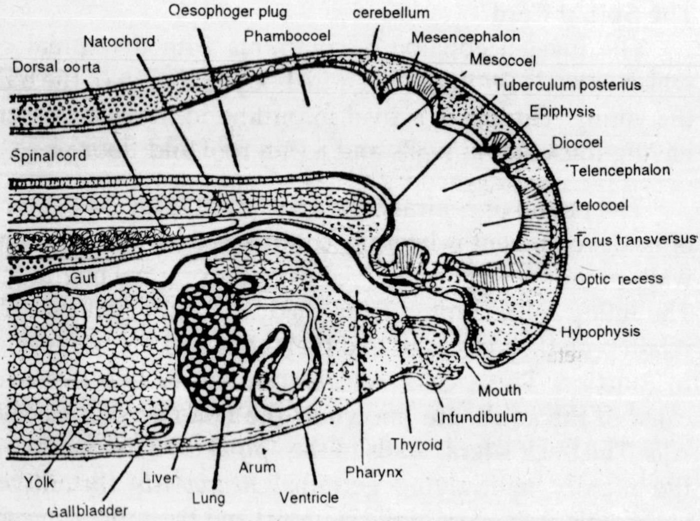

Fig. 5.5. Development of the brain and anterior structures of the frog tadpole. *(Top)* Medium sagittal section of the 7 mm. frog tadpole. *(Bottom)* median sagittal section of the 11 mm. frog tadpole.

The compact glia and neuroblasts close to the inner layer of ependyma comprise the *gray matter* of the cord. It is within this layer that the bulk of the cell bodies of neurons and the commissural fibers from one side of the cord to the other will be seen. As the cord develops further, the anterior and posteriorly directed fibers of the various neurons are concentrated more laterally so that only cross sections of axons will be seen. This peripheral area is then known as the *white matter*, to distinguish it from the more central and cellular gray matter of the cord.

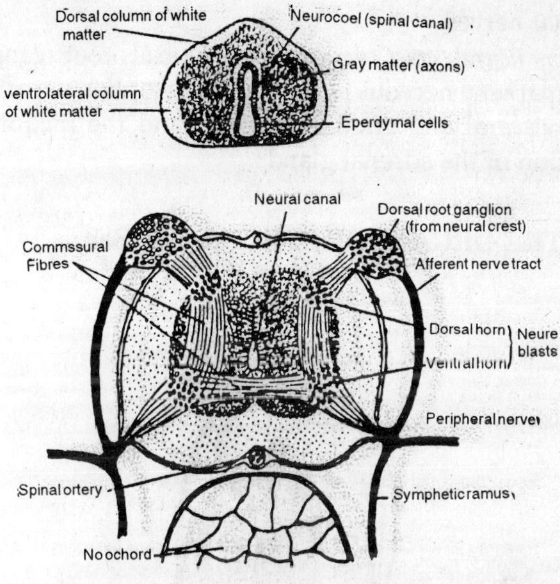

Fig. 5.6. Development of the spinal cord of the frog. *(Top)* Spinal cord of the 7 mm. larva. *(Bottom)* Spinal cord just before metamorphosis.

In addition to the lateral walls of the cord, which are thick, the dorsal wall is also thick because it consists of the fused neural folds. The central canal or *neurocoel* is therefore displaced ventrally in the embryo and larva, but, as the cord is developed and more neuroblasts are formed, the neurocoel assumes a more central position. Within the enveloping connective tissue membrane which surrounds the spinal cord may be seen several blood vessels, principally the large *spinal artery* located in the mid-ventral (inverted) groove of the cord.

The Peripheral Nervous System

The development of this variegated portion of the nervous system will be treated in the following sequence, and the description will be continuous to the final stage of development:

1. *Organs of speciial sense* : Optic, auditory, olfactory, and lateral line organs.
2. Cranial nerves I to X.
3. Spinal nerves.
4. *Other neural crest derivatives* : Dorsal root ganglia, sympathetic nervous system, chromatophores, parts of the visceral and cranial cartilage, and the medullary portion of the adrenal glands.

THE ORGANS OF SPECIAL SENSE

The Eye

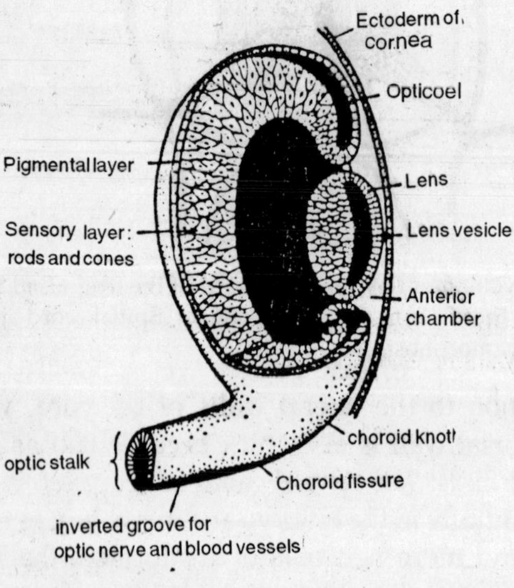

Fig. 5.7. Schematic diagram of the developing eye parts of the frog.

The *optic vesicles* arise early, by the tail-bud stage, as lateral diverticula from the ventro-lateral walls of the diocoel. The connection of the brain cavity with the optic vesicles become constricted, by the convergence of the sorrounding mesenchyme, into a tube known as the *optic stalk.*

Development of Eye

The development of eye is a complex morphogenetic powers. Its various tissues are derived from three source—the wall of the diencephalon (neural ectoderm), the superficial ectoderm (epidermal ectoderm) and the mesenchyme (mesoderm) of the head. During its morphogenesis, each of its component parts passes through the *phase of induction and rudiment formation; phase of cytodifferentiation* and *the phase of growth.*

(i) Development and differentiation of retina. Soon after the differentiation of three primary brain regions from the neural tube, the rudiments of eye begin to shape in the region of future diencephalon. The presumptive retinal plates which lie at both lateral sides of forebrain (diencephalon) begin to evaginate as a pair of sac-like protrusions of neural epithelium, called *primary optic vesicles.*

Each primary optic vesicle continues to grow and push outward by displacing the intervening mesenchyme until it comes in direct contact with the inner surface of lateral head epidermis. Its proximal end eventually becomes constricted from the remainder of diencephalon and remains connected with the latter by a narrow *optic stalk.* Now, the external wall of the optic vesicle which has contacted with the epidermis, flattens out and begins to invaginate to transform the primary optic vesicle into a double walled *optic cup* or the *secondary optic vesicle*, enclosing a new space, the future posterior chamber of the eye. Of the layers of optic cup, the pigment granules develop in the external thin layer which forms the *pigmented coat of retina.* In the invaginated thicker layer sensory and conducting cells differentiate to form the *neuro-sensory layer of retina*. These two layers are continuous with one another at the rim of the optic cup which forms the edges of the *pupil*. The pupil is quite wide in the beginning, but later on, its edges bend inward and coverge so that the opening of pupil reduces to its final dimensions. Later on, the edges surrounding the pupil become considerably thin to form the *iris*.

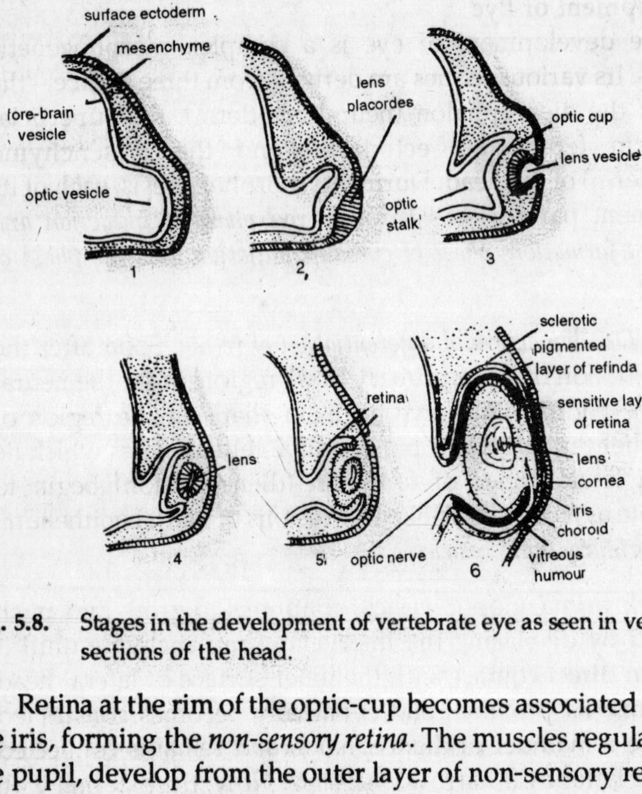

Fig. 5.8. Stages in the development of vertebrate eye as seen in vertical sections of the head.

Retina at the rim of the optic-cup becomes associated with the iris, forming the *non-sensory retina*. The muscles regulating the pupil, develop from the outer layer of non-sensory retina. Parts of the mesodermal choroid give rise to *ciliary body*. Some parts of this body are derived from the non-sensory retina.

Following invagination of the optic-cup, there is an increase both in the thickness and the area of developing neuro-sensory retina. The innermost neural layers of retina formed of ganglionic cell (nearest to the centre of eye), differentiate first and the sensory layer of rods and cones last. Axonic nerve fibres, leaving the neural layer (ganglionic cells) of retina, converge towards the optic stalk and grow through the wall of the stalk as *optic nerve.*

The peripherally directed dendritic processes of ganglionic cells, lining the innermost layer of neurons in the retina, form synapses with axons to form the outermost layer of neural retina called the *nuclear layer*. Blood vessels supplying the retina and other structures, enter the secondary optic vesicle through a gap, the *choroid fissure,* lying ventral to the optic stalk. This fissure closes during late embryonic stages.

The light-sensitive rods and cones develop from the outer cells of nuclear layer. Each cell producing a cytoplasmic bud at its outer end, which protrudes through a pore in the external limiting membrane. Further differentiation of rods and cones is sequential and centrifugal. The inner segments of rods and cones are first formed, and form them the outer segments later arise. Several specialized cell organelles, such as myoid, paraboid and oil drops develop in the inner segment. *Diplosome* (basal body) also migrates into the inner segment early in its development.

(*a*) The *myoid* is a acetylcholinesterase rich contractile region in the basal portion of the inner segment which in adult positions the photosensitive elements relative to the pigmented retinal epithelium during photochemical adjustment to different light intensities.

(*b*) The *ellipsoid or lentiform body* is a tightly packed aggregation of mitochondria.

(*c*) The *paraboid* is a dense collection of glycogen, derived from the granular endoplasmic reticulum.

(*d*) The *oil-droplets* are spherical globules of carotenoids, usually remain confined to the cones. They appear late in development between the mitochondria of ellipsoid and the tip of the inner segment. These droplets serve as focusing lenses as well the optical filters.

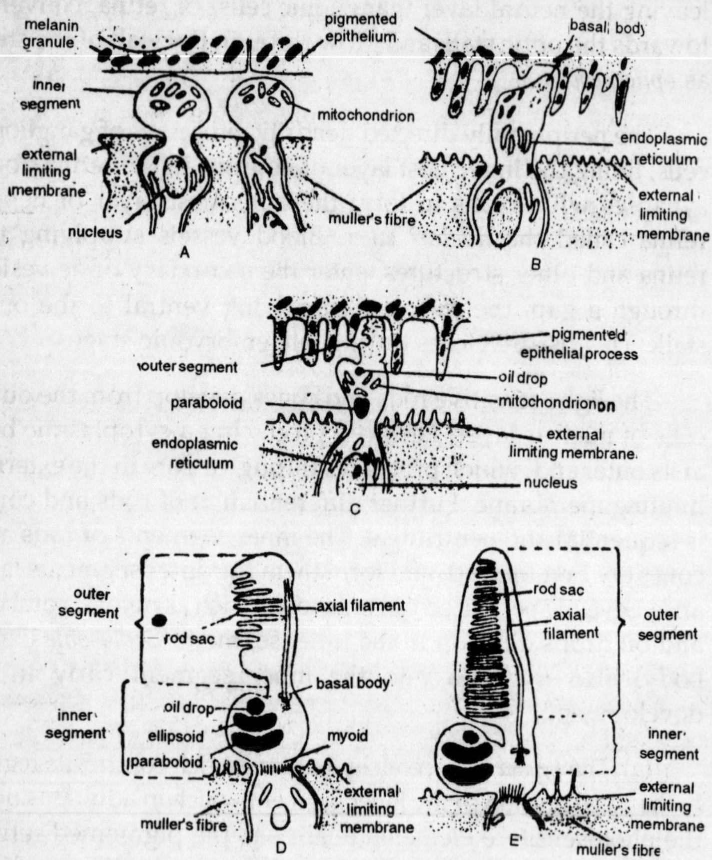

Fig. 5.9. Successive stages in the formation and maturation of rod and cone.

(*e*) The outer centriole of the *diplosome* or the *basal body* develops a non-contractile cilium which participates in the development of the outer segments of rods and cones. Each cilium produces nine pairs of peripheral ciliary microtubules. The peripheral doublets lengthen and collectively form the axial filament of the outer segment. Minute rows of ciliary vesicles extend from the tip of the ciliary tubules to the apex of the developing outer segment. At the point where ciliary vesicles contact the plasma membrane, the cell surface folds inward to form large, irregular sacs. These rod or cone sacs become flattened in a plane perpendicular to long axis of the outer segment and become stacked in a manner of a stack of coins. As this stack grows, the outer segment elongates.

The rod-sac is the presumed location of the photosensitive pigment *porphyropsin* or *rhodopsin*. In some amphibians, porphyropsin is preser before metamorphosis and is replaced by rhodopsin during metamo phosis.

In the meantime, the cuboidal epithelial cells of the outer wall of optic cup accumulate the granules of *melanin* pigment which are manufactured by the Golgi-complex of these cells. Soon each pigmented epithelial cell gives rise to numerous long, fine cytoplasmic processes over its inner surface. These intrdigitate with the outer segments of the rods and cones which arise simultaneously and thus, locking the two layers together.

(ii) Development of lens. Meanwhile, the prospective neuro-sensory retinal layer of optic vesicle inducts the epidermis of the head, opposite to the optic vesicle, which thickens to form the *lens-placodes.* In the later development, the lens placodes of each side invaginate to form the *lens cup* which gradually deepens and progressively pinches off, from the epidermis, until it forms a closed *lens vesicle.*

The lens vesicle comes to lie in the secondary optic vesicle. Cells forming the lens vesicle undergo cytodifferentiation or metamorphosis before it can function as a refracting body, the *lens.* During cytodifferentiation, the cells on the inner side of the lens vesicle elongate, become columnar and later on transform into long fibres of keratinous nature. During this transformation, the nuclei of cells degenerate and the cytoplasm becomes hard and transparent. The fibres become arranged in the lens in a very orderly manner, forming the spherical or ellipsoid refractile lens. The epithelial cells are continuously transformed into lens fibres, so that the refracting body grows by the apposition of new fibres.

(iii) Development of cornea. The cornea is formed by the skin epidermis and the mesenchyme of the head region. In larval stage, the epidermis forms the external cornea, which remains continuous with the skin, while the mesenchyme forms the internal cornea. The latter is in continuation with the sclera. During metamorphosis, these layers fuse under the influence of thyroid hormone and become transparent, so that the rays of light may enter the eye. The fully developed cornea consists of three kinds of cells, nerve fibres, a small number of collagen fibrils and mucopolysaccharides which become deposited in the interfibrillar space. The components of cornea develop a number of functional characteristics such as avascularity, tensile-strength, regenerative property, a characteristic interferance pattern in polarized light, an appropriate refractive index, a controlled curvature that contributes to its refractory power and transparency.

The space between the lens vesicle and the cornea contains a cellular material called *anterior vitreous body* or *aqueous humour*.

(iv) Development of choroid coat and sclera. During the course of development, the choroid coat and sclera of eye develop from the mesenchyme cells accumulating around the eye-ball. The inner layer of mesenchyme give rise to a network to blood vessels surrounding the pigmented retinal epithelium and is called *choroid coat*. The outer layer of mesenchyme forms a fibrous capsule, the *sclerotic coat or sclera*, around the eye. The sclera provides protection to eye and its muscles.

The *eye-muscles* are derived from the three pro-otic myotomes in the head. The premandibular myotome give rise to *superior, inferior* and *internal recti* and the *inferior obique muscles*. The mandibular myotome give rise to *superior oblique muscles* and the *external rectus muscles* is derived from the third, the hyoid myotome.

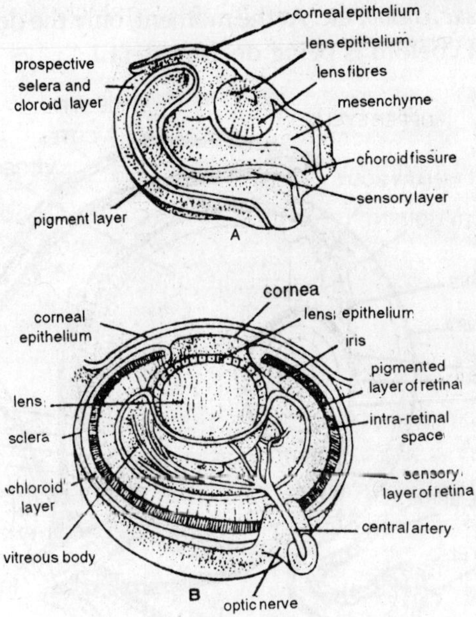

Fig. 5.10. Stereograms of developing eye

The two *eye-lids* and the *nictitating membrane* develop at a later stage. The inner lining of epithelium of the eye-lids forms the *conjunctiva* which is fused with the cornea.

It has been proved that a chain of inductions are operative during the development of eye. The ectoderm of the head region will not form the lens placodes unless it is not touched by the optic vesicle. Hence, the optic vesicle acts as *primary inductor* or *primary organizer*. Similarly, the lens placode acts as *secondary inductor*. It induces the neuro-sensory layer of optic vesicle to invaginate and when develops into lens, the latter induces the epidermis and mesenchyme for the formation of cornea, choroid coat and sclera.

Development of Mesodermal Organs

Like ectoderm the mesoderm is a complex rudiment of various important body organs such as *limbs, endoskeleton, heart, blood-vessels, kidneys, reproductive organs,* all kinds of *connective tissue, coelom* etc. At the moment, only the development of heart and coelom is being dealt in detail.

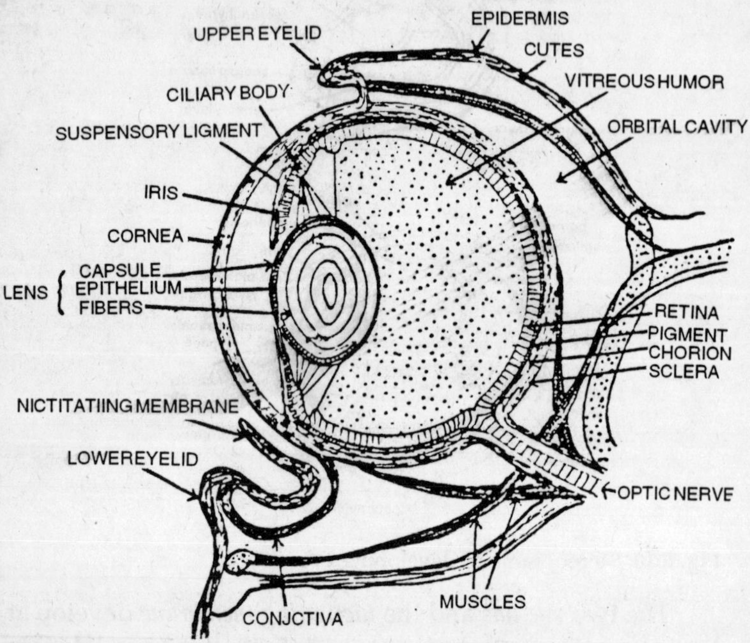

Fig. 5.11. Schematic Diagram through Frog's Eye

The Ear

The frog has no outer ear. The inner and the middle ear are developed much in the manner of all vertebrate ears, but to a less efficient and complicated degree.

The Inner Ear. The superficially placed *auditory placode* develops from nervous ectoderm on the side of the head at the level of the rhombencephalon, prior to the time of hatching (at about the 2.5 mm. stage). This occurs under the inductive influences of the medulla and archenteric roof. It moves inwardly toward the brain and invaginates to form a vesicle with but a temporary external opening at the 3mm. stage. This vesicle is the otocyst or *auditory sac.* After the head ectoderm closes over this invaginating vesicle at the 4 mm. stage, and the walls

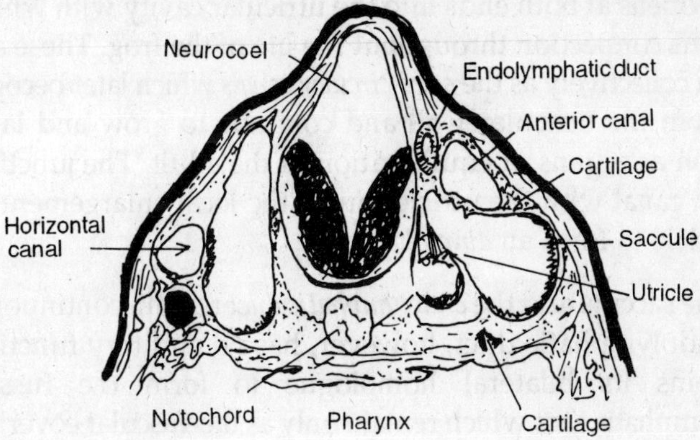

Fig. 5.12. Auditory apparatus of an 11 mm. frog tadpole.

become continuous, there develops from its dorsomesial wall a small evagination. This soon (at the 7mm. stage) becomes tubular and is known as the *endolymphatic duct*. The duct on each side grows dorsally to fuse with the duct of the other side of the brain. It finally loses its cavity and forms a vascular membrane which covers the rhombencephalon, having no auditory function whatever. The duct remains as a vestige even in the adult frog, originating between the membranous labyrinth (inner ear) and the hindbrain.

At about the 11 to 12 mm. stage there develops a vertical fold which divides the main cavity of the otocyst into mesial and lateral chambers. The more dorsal and mesial portion is the *utricle* and the more ventral and lateral portion is the *saccule*, both being joined by a small pore.

The utricle, by the 15 mm. stage, becomes further subdivided by there folds or ridges. These develop on the inside of the utricular wall. One is vertical and anterior *(anterior semi-circular canal)*, one is horizontal and lateral *(horizontal semi-circular canal)*, and finally there will appear a third fold which is vertical and posterior *(posterior semi-circular canal)*. These ridges fuse with one another to form three loop-like tubes, each of which opens at both ends into the utricular cavity with which it retains connection throughout the life of the frog. These are known collectively as the *semi-circular canals* which later become free from the utricular wall and continue to grow and later function as organs of equilibration in the adult. The junction of each canal with the utricle causes the local enlargement of the utricle to form an *ampulla*.

The saccule and the *endolymphatic duct* remain continuous. The endolymphatic duct, however, has no auditory function but joins its bilateral homologue to form the fussed endolymphatic sacs which remain only as the vascular covering of the hindbrain. The saccule, however, gives rise (at the 15 mm. stage) to two evaginations or sacs. One of these is posterior and ventral, known as the *cochlea* or *lagena*, and the other is slightly more dorsal (but psteriorly directed) and is known as the *pars basilaris* (basilar chamber) of the ear. These auditory structures never develop to the extent that they do in higher vertebrates. However, in all vertebrates the lining epithelial walls of the utricle (1), saccule (1), cochlea (3), and the various ampullae (3) develop sensory patches (as numbered) which are connected functionally with the eighth cranial or the auditory nerve.

The sensory (auditory and equilibratory) parts of the ear are therefore purely ectodermal in origin. However, the entire ear is end cased in mesenchyme which gives rise to cartilage that is finally displaced by bone. This makes up the *auditory capsule* which is mesodermal. Between the capsule and the inner ear (membranous labyrinth) is the *perilymph space* which is filled with the perilymph fluid, derived from mesenchyme.

The Middle Ear. This portion of the ear grows as an endodermally lined tube extending dorso-laterally from the original hyomandibular pouch and developing a terminal chamber. The chamber comes to lie between the inner ear and the head ectoderm. That portion of the endoderm in contact with the head ectoderm (original hyomandibular plate) becomes

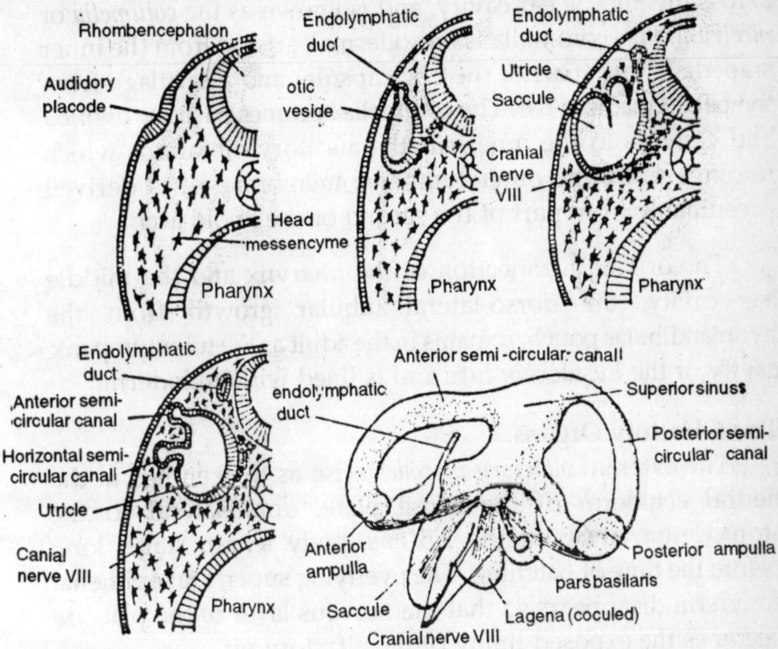

Fig. 5.13. Development of the auditory apparatus of the frog.

the *tympanic membrane* (ear drum), later to be invaded by mesenchyme which forms blood vessels and connective tissue. This membrane is supported by a cartilaginous ring known as the *annulus tympanicus* (tympanic ring) which is believed to arise from the palatoquadrate bone. These latter two structures are presumed to cause the development of the tympanic membrane by induction.

This endodermally lined chamber expands inwardly to make contact with the inner ear capsule at the site of the *foramen ovale,* an aperture in the inner ear capsule and extending through the perilymph space. This aperture is obstructed by a cartilaginous plug known as the *operculum.* The operculum later becomes ossified into bone. A cartilaginous rod arises from the roof (dorsal wall) of this cavity and extends from the operculum of the inner ear to the tympanic membrane and across the middle ear cavity, and is known as the *columella* or *plectrum*. This columella is mesodermal, arising from the inner stapedial plate (part of the otic capsule) and a cartilage from the palatoquadrate bar. The columella becomes partially ossified and conveys to the inner ear the auditory vibrations which impinge upon the outer tympanic membrane. It is derived presumably from part of the second or the hyoid arch.

The original connection of the pharynx and the middle ear cavity, the dorso-lateral tubular growth from the hyomandibular pouch, remains in the adult as the tubo-tympanic cavity or the *Eustachian tube* and is lined with endoderm.

The Olfactory Organs.

The external *olfactory placodes* arise as thickenings in the neural ectoderm of the sense plate, dorso-lateral to the stomodeum, at about the 2.5 *mm.* body length stage, long before the time of hatching. The overlying superficial epithelial ectoderm disappears so that the nervous layer of the placode becomes the exposed lining of the olfactory pit.

After hatching (6 mm. stage) a solid rod of ectodermal cells grows ventro-laterally from the olfactory pit to become attached to the pharynx just dorsal to the oral plate (i.e., stomodeum). By the 11 mm. stage this core of cells acquires a lumen which is continuous from the external nares *(olfactory pits)* to the internal nares *(internal choanae)* which open into the pharynx. The major part of this olfactory tube is lined with non-sensory epithelium. It develops a dorsal and a ventral chamber (sacs), each of which acquires glandular masses which are known as the organs of *Jacobson*.

The neuroblasts of the olfactory placodes send extensions posteriorly and give rise to the fibers which form the *olfactory* or first cranial nerve. These grow toward the brain and are guided in their directional development by the outgrowth of the telencephalon known as the *olfactory lobes*.

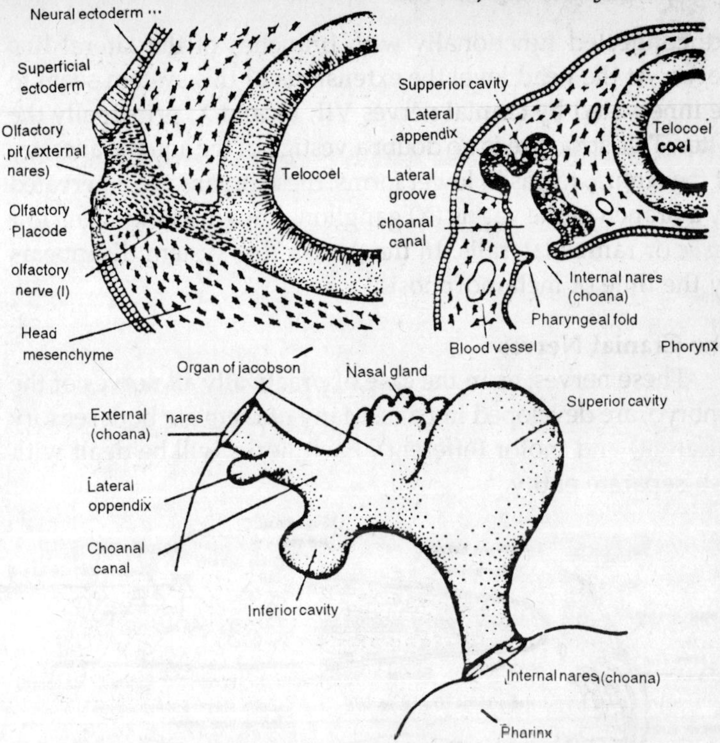

Fig. 5.14. Development of the olfactory organ of the frog. *(Top, left)* Sagittal section through the olfactory placode and nerve. *(Top, right)* Posterior transverse section through the choanal canal. *(Bottom)* Schematic reconstruction of the embryonic olfactory organ.

The Lateral Lines Organs

The most posterior or fourth cranial placode (cranial nerve X) sends a growth posteriorly beneath the lateral body epidermis, on either side, beginning at about the 4 mm. stage. It grows posteriorly to the tip of the tail. Along these cords arise groups of sensory cells which grow through the epidermis to become exposed along the sides of the body as the *lateral line system*. The exposed cells are ciliated, and are therefore sensitive to vibration in the surrounding aquatic medium. They are protected by inner and outer sheath cells and a basement membrane and are connected functionally with branches of the lateral line nerve. At the head level the extensions of this system seem to be innervated by cranial nerves VII, IX and X, principally the latter. This structure is no doubt a vestige of the aquatic ancestry of the Anura, for, as in lower forms, these organs are innervated by a branch of the vagus (X) ganglion, known as the *lateral line nerve* or ramus lateralis. In the Anura this system disappears by the time of metamorphosis.

The Cranial Nerves

These nerves, as in the case of practically all nerves of the embryo, are developed in pairs. Many of them are both sensory (afferent) and motor (efferent). Each nerve will be dealt with as a separate entity.

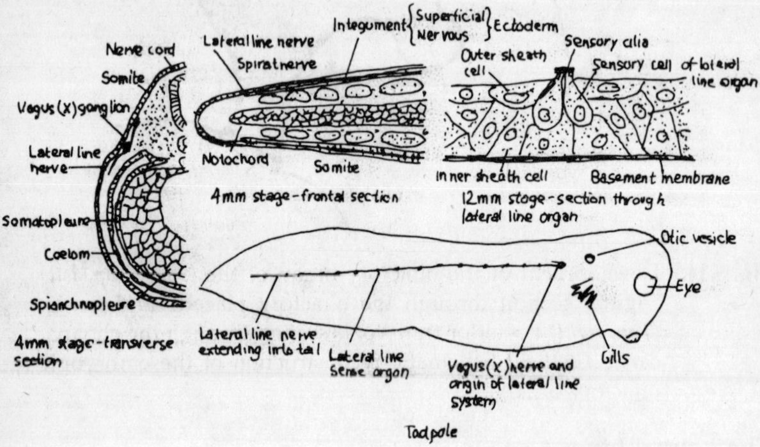

Fig. 5.15. Origin of the lateral line sense organ system in the frog larva.

As the neural folds close they leave a column of cells between the fold and the dorsal ectoderm. These are known as the *cranial* or *neural crests*, depending upon the level under consideration. The *cranial crests* of the brain level are different from the neural crests of the spinal cord level in that they become divided into four large pairs known as the *cranial placodes*. These consist of thickened patches of neural ectoderm of the head, one of which, the auditory placode, has been described already. The original cellular connection between each cranial crest segment and the brain remains as the sheath of the nerve fibres to be developed in that region. At the spinal cord level the segments are metameric and will be as abundant as the spinal nerves developed therefrom.

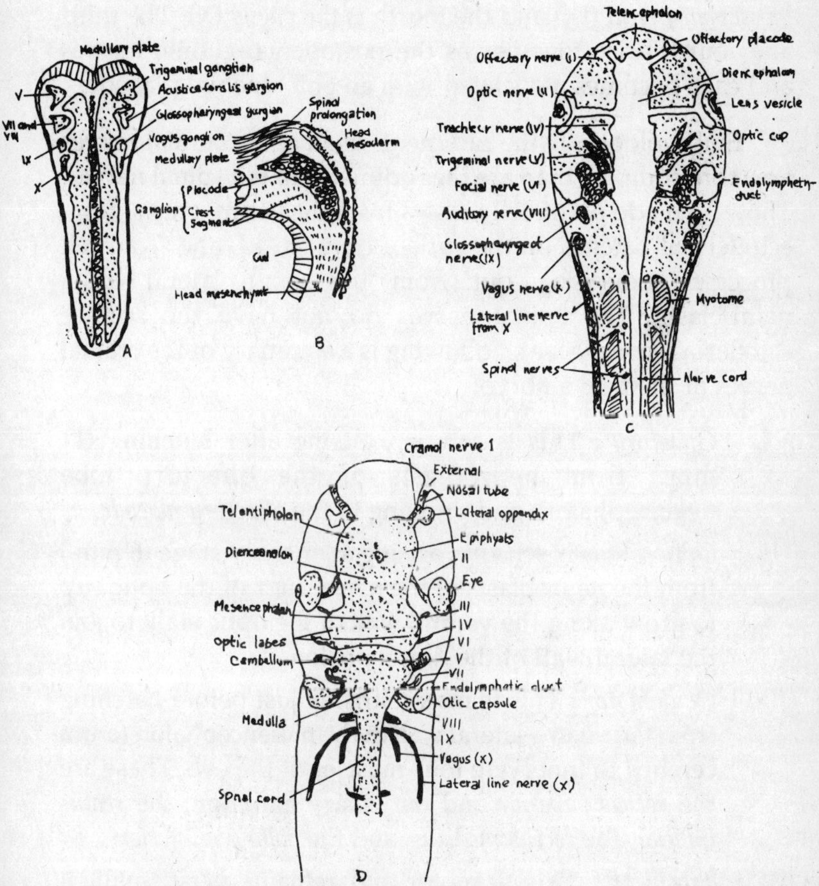

Fig. 5.16. Origin and derivatives of the cranial ganglia. (A) Origin of the cranial ganglia: frontal section. (B) Relation of the crest segments and placodes: transverse section. (C) Relation of the sense organs to the cranial ganglia. (D) Brain, sense organs, and cranial nerves of the 12 mm. tadpole.

Four ganglionic masses are found in the development of the frog cranial nerves as early as the 6 mm. stage. These include the first or *semilunar ganglion* of the fifth nerve. The second is the *acusticofacialis* (VII-VIII) ganglion which becomes associated with the auditory placode. The third is the *glossopharyngeal* (IX) and the fourth is the *vagus* (X). The third and fourth arise together as the postotic ganglionic masses and each becomes associated with an epibranchial placode.

Embryologically the rudiments of all of the cranial nerves are more complex than are the rudiments of the spinal nerves. They are made up of cell masses from the neural crests, from ectodermal patches on the surface of the head, and from cell processes extending out from the ventro-lateral brain neuroblasts. The spinal nerves do not have the surface ectodermal constituent. Following is a summary of the cranial nerves of the frog embryo:

I. *Olfactory* : This is sensory, arising after hatching (11 mm.) from neuroblasts of the olfactory lobe (telencephalon) and growing to the *olfactory placode.*

II. *Optic* : This is sensory, arising at an early stage (6 mm.) from the neuroblasts of the *retinal layer* of the optic cup to grow along the ventral wall of the optic stalk to join the lateral wall of the diencephalon.

III. *Oculomotor* : This is motor, arising just before hatching from the ventro-lateral wall of the mesencephalon (crura cerebri) to innervate four muscles of the eye. These are the *rectus superior* and the ciliary ganglion, the *rectus inferior*, the *rectus medialis*, and the *obliquus inferior.*

IV. *Trochlear* : This is motor and remains very small. It arises late in development from near the roof of the mesencephalon between the optic lobes and the cerebellum just posterior to the valvula cerebelli and the posterior commissure. It innervates the *superior oblique muscles* of the eye.

Segments of neural crest	I	II	III	IV	V	VI
Nerves derived from them	V ramus ophthal-micus	V main branch	VII and VIII	IX	X	dorsal root of 1st spinal

	Head Somites			Trunk Somites			
Meso-dermal Somites	Preman-dibular	Man-dibular	Hyoid	1st	2nd	3rd	4th
Ventral roots	I	II	III	IV	V	VI	
Nerves derived from them	III	IV	VI	dis-appears	dis-appears	ventral root of 1st spinal	

(Courtesy, Jenkinson: "Vertebrate Embryology," Oxford, The Clarendon Press).

V. *Trigeminal* : This is a mixed nerve, also known as the trifacial ganglion, which appears as early as the 5 mm. stage and has a complicated origin from the dorsal (superficial) ganglionic elements of the first cranial crest segment and the ganglionic portion of the first cranial placode. It is well-developed by the 9 mm. stage. The level is at the anterior end of the medulla. This is a very large and tri-radiate ganglion, including an *Ophthalmic* branch from the placode whose fibers pass to the skin and snout, land a forked *Gasserian ganglion* which arises from both the crest and placode elements and gives rise to the *maxillary* (upper) and *mandibular* (lower) branches of the fifth nerve. The entire mixed trigeminal nerve ganglion (V) joins the dorso-lateral wall of the medulla oblongata. The outer or non-nervous portion of the placode disappears and it is believed that the non-nervous portion of the crest segment may contribute some of the mesenchyme to the mandibular arch. It is certainly difficult to distinguish between them.

VI. *Abducens* : This is a motor nerve arising late (11 mm

stage) from the neuroblasts of the ventral side of the medulla and innervating the lateral or external rectus and *retractor bulbi muscles* of the eye.

VII. *Facial* : This is a mixed nerve, arising by the 5 mm. stage from the anterior ganglionic portion of the second or *acusticofacialis* (second cranial) *crest segment* and *placode* and is associated with neuroblasts coming from the medulla just posterior to cranial nerve V. The motor (efferent) fibers are divided between the *hyoid* and *palatine* (mouth) branches, but a good portion of the crest segment, from which these fibers arise, is non-nervous and is presumed to give rise to *mesenchyme* of the hyoid arch.

VIII. *Auditory* : This is a sensory nerve arising by the 5 mm. stage from the posterior ganglionic portion of the second *(acustico-facialis)* crest ganglion, in close association with the auditory placode, and innervates only the inner ear.

IX. *Glossopharyngeal* : This is a mixed ganglion, arising by the 9 mm. stage from the ganglionic portion of the third *cranial crest segment* and *placode*. Fibers of the ninth and tenth cranial ganglia enter the sides of the medulla together, but peripherally the ninth or glossopharyngeal nerve supplies the *first branchial arch*, the *mouth*, the *tongue*, and the *pharynx*. This ganglion is incompletely separated from the vagus by the anterior cardinal vein.

X. *Vagus* or pneumogastric ganglion : This is mixed and arises by the 9 mm. stage from the fourth cranial crest segment and placode and becomes associated with the neuroblasts from the medulla. Peripherally the branches feed the *second* and *third branchial arches*, the *tympanum*, the *muscles* of the *shoulders*, the *viscera* (including larynx, oesophagus, stomach, lungs and heart), and the lateral line system. The non-nervous portion of the crest segment forms mesenchyme and the superficial non-nervous elements of the placode then disappear.

Summary of Derivatives of the Cranial Nerves in Tabular Form

Cranial Crest Segment	*Cranial Nerves Derived Therefrom*
I	V ophthalmic branch
	V mandibular branch
II	VII and VIII
III	IX
IV	X

The Spinal Nerves

The *spinal nerves,* aside from their component parts, differ from the cranial nerves in that they are related to mesodermal somites rather than visceral clefts.

The spinal nerves arise from the pair of continuous *neural crests.* These are elongated strands of neural ectoderm left outside of the dorsal neural tube, between it and the dorsal ectoderm, at the time of closure of the neural tube. These crests become metamerically segmented and are paired as a result of the development of the somites. In frontal sections of the 7 mm. stage these crests will be seen as dorsal root ganglia of the spinal nerves. No placodes are developed in association with any part of the neural crests, as were described in association with some of the homologous cranial crests.

Each crest segment is made up of many neuroblasts which send fibers to the dorso-lateral wall of the spinal cord to form the *dorsal root* or ramus of the spinal nerve. This connects the *dorsal root ganglion* (original neural crest) with the spinal cord. Also, other fibers grow ventro-laterally from this ganglionic crest to become the afferent tracts from the skin and other sensory organs.

Within the spinal cord, as early as the 4 mm stage neuroblasts develop and send out bundles of efferent axons (fibers) which emerge from the ventro-lateral wall of the spinal cord and almost immediately co-mingle with the afferent fibers, already described. The *ventral nerve roots* or rami and entirely efferent, the *dorsal nerve roots* are entirely afferent, but the nerve trunks contain, within a common sheath, both the afferent and efferent fiber trunks. Many of these spinal nerve trunks fork to send both types of fibers dorsally to the sense organs and ventro-laterally to the muscles and to the limbs.

The tadpole may have a total of 40 or more pairs of spinal nerves, but only the anterior 10 pairs remain after the degeneration of the tail lat metamorphosis. The first pair *(hypoglossal)* emerges from between the first and the second vertebrae, to innervate the tongue and some of the muscles of the hyoid arch. The second pair *(brachial)* emerges from between the second and the third vertebrae and are very large. They have connections with the first and the third spinal nerves and thus form the large *brachial plexus,* innervating the forelimbs and the muscles of the back. The third pair has the aforementioned connection with the brachial but it then supplies the external oblique and transverse muscles and the skin. The fourth, fifth and sixth spinal nerves supply the skin and muscles of the abdominal wall. The seventh, eighth, and ninth spinal nerves together form the *lumbosacral* or *sciatic plexus* which innervates the posterior abdominal region and the hind legs. The tenth, with a branch from the ninth, forms the *ischio-coccygeal plexus* which sends branches to the urinary bladder, cloaca, genital ducts, and posterior lymph hears.

THE NEURAL CREST DERIVATIVES

The Sympathetic Nervous System.

The development of the sympathetic nervous system is not understood clearly but it is believed also to arise from the neural crests, by a downgrowth along the afferent trunk to the spinal nerve and thence to the dorsal aorta and the viscera beginning at about the 6 mm. stage. The sheath cells probably come from the neural tube itself. Just before hatching there may be seen clusters of cells around the junction of the dorsal and ventral nerve roots, and these clusters will give rise to the *sympathetic ganglia* which become associated with the *dorsal aorta* and the *viscera*. These ganglia are also interconnected by paired longitudinal strands on either side of the spinal cord. The fibers which join the sympathetic ganglia with the spinal nerve are known collectively as the *ramus communicans*.

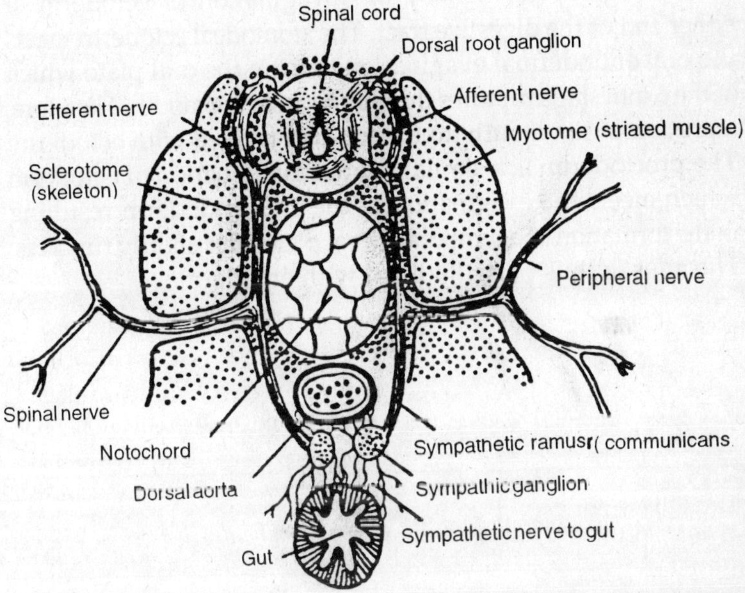

Fig. 5.17. Relation of the spinal and the sympathetic nervous systems.

The sympathetic system arise with the vagus (cranial nerve X), and its connecting fibers extend through the jugular foramen of the skull and along each side of the spinal cord to receive these connecting rami from the segmental spinal nerves. Both independent sympathetic nerves and plexi are developed and are found often quite removed from their source or origin.

The Adrenal Medula

The large brown cells of the inner portion of this endocrine organ arise sometime before metamorphosis by migration from the sympathetic ganglia and hence are of ectodermal origin.

They have lighter staining nuclei and are scattered among the more abundant cortical cells. These latter cells arise from the mesoderm and will be discussed subsequently.

The Stomodeum And Proctodeum

The *stomodeum* and *proctodeum* are also of ectodermal origin, each being represented by an invagination of ectoderm at either end of the digestive tract. The stomodeal ectoderm meets the oral endodermal evagination to form the oral plate which at the 6 mm stage ruptures through to form the mouth. Therefore, a portion of the mouth (to the tongue) is lined with ectoderm. The proctodeum is a similar caudal invagination of ectoderm which meets the evagination of the anal endoderm resulting in the formation of a temporary anal plate. This shortly ruptures. The anus is thus partially lined with entoderm.

6

The Mesodermal Derivatives

THE DIFFERENTIATION MESODERM

Mesoderm is either derived from embryonic ectoderm a from embryonic endoderm. In most of cases including frog it is developed from endoderm of the archauteron. During development the mesodermal part is given and in the fair of mesodermal pouches which soon differentiate into *upper mesodermal somites* and lower *splanchnic mesoderm*. The mesodermal somites are arranged metamerically from anterior to posterior.

The first ones to be completed are located just posterior to the auditory capsule. A total of 13 pairs is formed from the head to the base of the tail. In the tail of the 6 mm. tadpole there may develop as many as 32 pairs of these transient somites, making a total of 45 pairs by the 6 mm stage. The two most anterior or occipital pairs of somites become indistinguishable and the larval tail somites are lost during the process of metamorphosis. This leaves a total of 11 pairs of somites from which develop most of the striated muscles and the primary skeleton of the body. In the head and pharyngeal region the mesoderm is in the form of loose mesenchyme.

Each somite develops in a characteristic manner, having an outer thin layer of cells known as the *dermatome* or *cutis plate* and a mesial mass of cells known as the *myotome* or *muscle segment*. Between there is a cavity, the *myocoel*, which becomes elongated dorso-ventrally as the cutis plate extends to the body wall and appendages. Originally this myocoel is continuous with the splanchnocoel or body cavity. The myocoel eventually becomes obliterated. From the mesial margin of the somite, *sclerotomal cells* are proliferated off to take up their positions as a continuous skeletogenous sheath around the spinal cord and notochord, and between the segmental myotomes. This embryonic sheath gives rise to the cartilage and finally the bone of the axial skeleton. A total of nine anterior well-formed vertebrae develop in this manner, alternating with the myotomes. A tenth vertebra is modified into a *urostyle* or posteriorly elongated bony projection. The sclerotome of the two potential posterior vertebrae fuses to form the cartilage and then the bony urostyle which encloses the posterior end of the notochord.

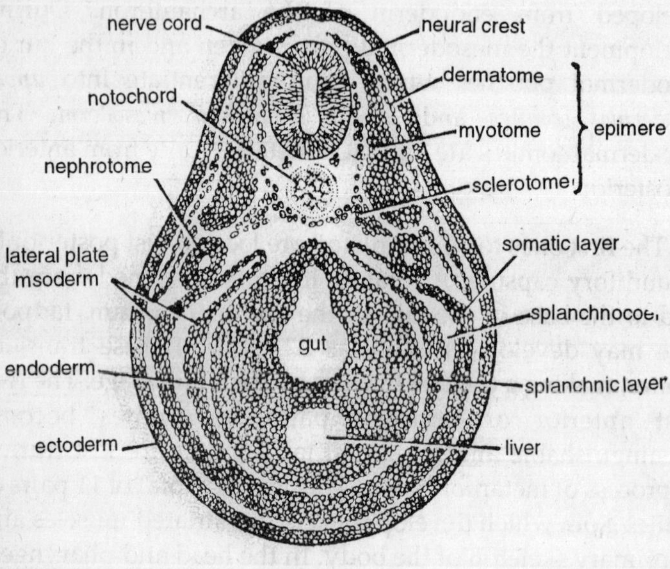

Fig. 6.1. T.S. young tadpole through trunk region showing differentiation of mesoderm.

The *dermatome* or *cutis plate* gives rise to the dermal layer of the dorsal and lateral skin, to connective tissue between the myotomes, and to the musculature of the limbs. The dermal layer of the ventral body wall arises from the somatopleure, so that while the dermis becomes continuous it originates from two sources.

The *myotomes* or *muscle segments* of the early larva arise as rather solid aggregations of cells, concentrated at the level of the spinal cord and notochord. They enlarge at the expense of the contained myocoels. Before the time of hatching, these myotomal cells become elongated and their muscle fibrillae orient their axes in the longitudinal direction of the embryo. This indicates a shift in the axes from the original direction in the early myotomes. The myotomes become separated from each other by septa or myocommata (connective tissue sheets). These are derived from the cutis plate, and assume a "V" shape with the apex of the "V" pointing posteriorly. These myotomes become the layer of striated or voluntary (voluntarily controlled) muscles of the back, limbs, and dorsal body wall. The fibers of many of these muscles are arranged ultimately in a variety of directions to provide for complete body control.

The muscles of the heart, blood vessels, and viscera arise from the hypomeric splanchnopleure and are known as smooth or involunary muscles.

Limb buds develop from the accumulation of loose mesenchyme of adjacent somites surrounded by ectoderm. Within this blastema of somatic mesoderm there arise muscle and bone, developing from the body outwardly. The forelimb muscles develop before those of the hindlimbs, although the hindlimbs are the first to emerge during the process of metamorphosis.

The mesodermal somite proliferate a number of cells — the *scleroblasts* arranged in loose groups called *scleroblasts*. The scleroblasts give rise to the entire endoskeleton including all cartilages and bones of the body.

The Vertebral Column

The development of the *notochord* has been described already. Its origin cannot be attributed exclusively to any of the three germ layers. Its cells remain few in number, become flattened antero-posteriorly, and are highly vacuolated. The notochord then becomes surrounded by (1) the primary or *elastic sheath* derived from the notochordal cells, (2) the secondary or *fibrous sheath,* and (3) eventually an outer *skeletogenous sheath* which is of mesodermal origin. This later connective tissue sheath encapsulates the nerve cord and also extends laterally between the myotomes by the 15 mm. stage. The notochord finally disappears in the frog. It is partially replaced by and partially converted into material of the centrum of each vertebra. Neural arches and transverse-processes develop outwardly from the cartilaginized skeletogenous sheath.

The *vertebral column* of the frog consists of ten vertebrae of which the last is modified into an elongated rod-like bone known as the urostyle, mentioned above. The first vertebra is the cervical *atlas,* followed by seven abdominal vertebrae and finally the ninth or *sacral* vertebrae. The *urostyle* takes the place of the caudal vertebra. The skeletogenous sheath from the sclerotome encloses the spinal cord and notochord by the time of hatching, and by the 15 mm stage some of this sheath has given rise to cartilage. It is within this cartilage that the vertebrae develop. The formation of this cartilage, due to the accumulation of the sclerotomal cells between the myotomes, is segmented in a manner which alternates with the developing muscle segments. However, at all levels the spinal cord and notochord will be enclosed in this double ring of connective tissue which is being transformed progressively into cartilage and finally into bone. Around the notochord it is known as the *perichondrium* and around the spinal cord as the *vertebral cartilaginous* arch.

Ossification begins, and the cartilage gradually is displaced by an inward growth of true bony cells known as *osteoblasts.* Eventually the notochord itself becomes invaded by these bone-forming cells and is transformed into the bony *centrum* of the vertebra. Each centrum becomes concave anteriorly, for the reception of the convex projection of the more anterior centrum. At the level of the lateral myotomal mass, connective tissue invades the notochord to form the *inter-vertebral discs* or *ligaments* of hyaline cartilage. Each disc splits into an anterior and posterior part, and each becomes ossified and later fuses with the corresponding part of the adjacent centrum. There are connecting ligaments which arise from the original sclerotome along both the dorsal and the ventral faces of the centra. They alternate in position with the myotomal segments.

The ossified cartilage which surrounds the spinal cord is known as the *neural arch*. Both the anterior and the posterior margins of each neural arch bear a pair of short zygapophyses which are processes that articulate and tend to join successive vertebrae. Spinal nerves emerge from the spinal cord through intervertebral foramina, between the sides of the succeeding neural arches.

Dorsal to each potential vertebra surrounding the spinal cord is a single median cone of ossification which becomes the spinous or neural process of the vertebra. The successive neural spines are connected by ligaments. The *paired transverse* and *bony processes* arise laterally, at right angles, from the cartilage of the centrum and become continuous with the minute bony ribs. No transverse processes or anterior zygapophyses are developed on the most anterior vertebra, the atlas. This vertebra is modified to articulate with the occipital condyles of the skull. The transverse processes of the ninth vertebra are elongated and are directed obliquely posterior to provide attachment for the ilium of the pelvic girdle. The tenth vertebra is greatly modified into the single tubular urostyle and is derived from the sclerotome (skeletogenous material) of the last two somites of the body.

The Skull (Neurocranium of the Adult)

The *skull* of the adult frog, defined as that part of the skeleton which surrounds and supports the brain and special sense organs, is composed of relatively more cartilage than is the skull of most higher vertebrates..The cranial cartilages are an exception to the general rule as to origin, and are derived from ectodermal neural crests. The frog skull represents a type intermediate between the skulls of lower fishes and higher reptiles. It is made up of the *cranium,* which arises from the cartilage and is therefore known as the *chondrocranium;* the *sense capsules;* a portion of the *notochord;* and a *visceral skeleton,* which includes the jaws and the hyoid apparatus and membrane and dermal bones. There are cartilage (endochondral) *bones* which arise by the ossification of cartilage and there are *membrane* (dermal) *bones* which develop from (dermis) superficial membranes without the intervention of a cartilage phase. These latter bones are sheet-like and can be stripped from the cartilage of the true cranium.

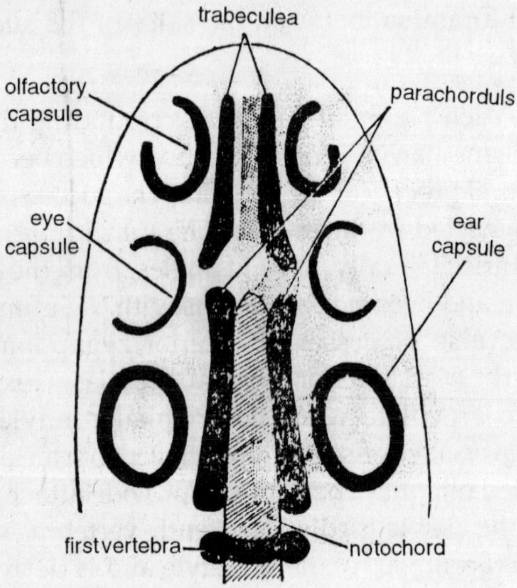

Fig. 6.2. Formation of component elements of the vertebrate chondrocranium.

The bony floor of the cranium arises largely from the notochord in conjunction with a pair of carilaginous (sclerotomal) rods known as the *parachordals* at about the 7 mm stage. These are arranged longitudinally on either side of the notochord with which they become fused to form the

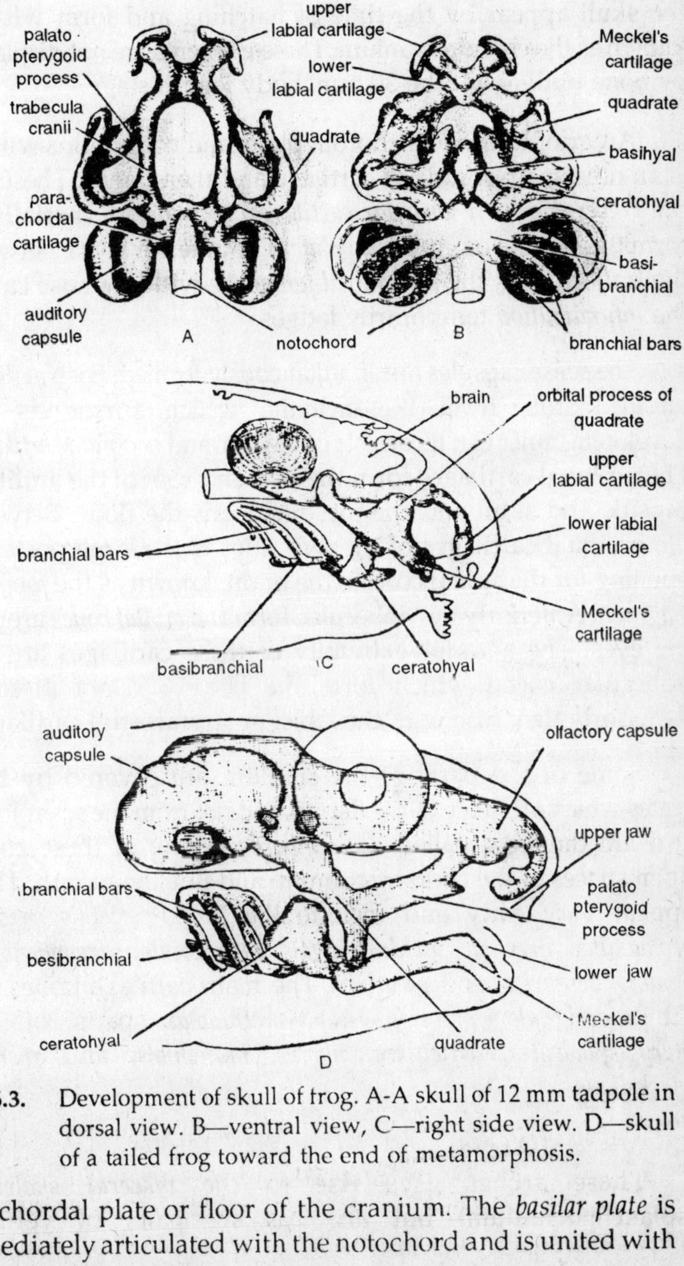

Fig. 6.3. Development of skull of frog. A-A skull of 12 mm tadpole in dorsal view. B—ventral view, C—right side view. D—skull of a tailed frog toward the end of metamorphosis.

parachordal plate or floor of the cranium. The *basilar plate* is immediately articulated with the notochord and is united with the parachordals on either side. The cartilaginous elements of the skull appear by the time of hatching and form what is known as the chondrocranium. These elements are not displaced by bone until about the 30 mm. body length stage.

Anterior to the parachordal plate, and continuous with it, is an other pair of rods of cartilage and then bones. These are the *trabeculae* (or *trabecular cartilages)* which join mesially to form the *ethmoid* or *intranasal plate*. The anterior space between the trabeculae is the *basicranial fontanelle* within whose cavity the *infundibulum* temporarily lodges.

The sense capsules are simultaneously formed. Each *auditory capsule* forms from mesenchyme which surrounds the developing inner ear. It consists of mesotic and occipital *cartilages*. The occipital cartilages form the side and roof of the auditory capsule and skull and the mesotic forms the floor. Between the occipital cartilages of the two sides is the large posterior opening for the spinal cord to the brain, known as the *foramen magnum*. Anteriorly the trabeculae form the *orbital bones* around the eyes. The anterior extremity of these cartilages are the trabecular cornu which form the bony *olfactory capsules*. Anteriorly they fuse with the labial or suprarostral cartilages.

Some of the parts of the cranium are covered by thin plates which originate in the dermis and not from the sclerotome and are therefore called *dermal bones*. Many of these cover open spaces of the chondrocranium and line the mouth. They appear very early and include the *fronto-parietals, nasals, premaxillae, maxillae, quadratojugals, squamosals, parasphenoids, vomers, dentaries,* and *palatines*. The main cartilage bones are the *exoccipitals, prootics, stapes, ethmoids,* parts of the *pterygoquadrates, articulars, hyoids, branchials,* and *mento-Meckelians.*

The Visceral Arches

These arches give rise to the *visceral skeleton* (splanchnocranium) but arise as six pairs of vertical condensations of mesenchyme lateral to the embryonic pharynx. After the mouth opens each of these arches develops cartilage. The most anterior pair, the mandibular arches, give rise to the dorsal *palatoquadrate bone* and to the ventral Meckel's cartilage. The *palatoquadrate* (pterygoquadrate) *cartilage* arises from the *maxillary* (dorsal) portion of this first visceral arch and joins the posterior end of the *trabeculae*. The palato portion gives rise to the bulk of the *upper jaw* and the quadrate portion to the *annulus tympanicus* of the middle ear. The *Meckel's cartilage* gives rise to the core of the lower jaw or the *dentary bone*. The large *ceratohyal* cartilages appear posterior to the Meckelian cartilage in the hyomandibular arch. The *basicranial* (copula) is an unpaired cartilage found between the two large ceratohyals. The more anterior *basihyal* is slow to develop cartilage but can be identified by the U-shaped thyroid gland which appears directly ventral to it. The jaws are suspended to the skull by a cartilaginous bone which develops from membranes.

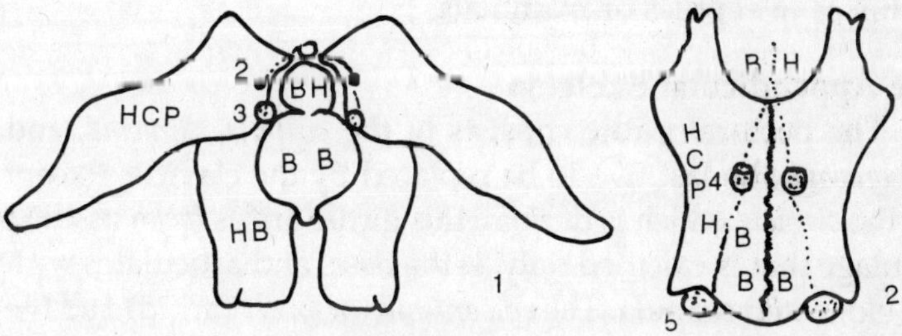

Fig. 6.4. Development of the thyroid gland and related hyoid cartilages. (1) Diagrammatic representation of the developing thyroid of the frog in relation to the hyoid cartilages in early metamorphosis; (HCP) hyoid cartilage proper, (BH) basihyoid cartilage; (BB) basibranchial cartilage, (HB) hyobranchial cartilage, (1) approximate initial side of the thyroid anlage, (2) approximate position of the thyroid at the hindlimb bud stage, (3) approximate position of the thyroid at the fully differentiated hindlimb bud stage.

(2) Same as "1" except at later stages of metamorphosis; (4) Approximate position of the thyroid at forelimb emergence, (5) approximate position of the thyroid at the tail resorption stage.

Portions of the second visceral (hyoid) arch give rise to the *hyoid cornu*. Parts of the third and fourth visceral arches (first and second branchials) give rise (at the 9 mm. stage) to the plate-like *hypobranchial* cartilage apparatus of the adult. The four pairs of slender *ceratobranchials* extend posteriorly from the hypobranchials. None of the arches remain as such by the time of metamorphosis, the only remnants being the derivatives just listed. The frog tadpole does not posses true teeth, but does have ectodermally covered horny "jaws" and so-called teeth. Oral papillae, appearing as teeth, constantly are wearing away and bear no relation to the bones with which the more functional teeth of the post-metamorphic frog will be associated. At metamorphosis dermal papillae appear on the upper jaw, associated with the vomers, and are known as tooth germs. The overlying epidermis is then involved in transforming the papillae into pseudo-teeth, with a minute amount of dentin and enamel. The teeth are for holding rather than for macerating living food, and never are developed as highly as in reptiles or mammals.

The Appendicular Skeleton

The pectoral girdle consists of the *scapula, coracoid*, and *precoracoid*, the last two to be replaced by the clavicle. Except for the *clavicle* which is dermal, this girdle arises from ossified cartilage that is calcified only at the base, and articulates with the elongated scapula. The *epicoracoid* (or precoracoid) and the *xiphisternum* remain as cartilage while the single *sternum*, paired coracoids, and single *episternum* become ossified. The sternum arises from the fusion of a longitudinal pair of cartilages which never join the ribs but remain both anterior and posterior to the pectoral girdle. The *humerus*, or upper portion of the forelimb, articulates with the *glenoid cavity* of the *pectoral girdle*.

The *pelvic girdle* is a V-shaped fused mass of bone which supports the hindlimbs and is associated with the urostyle. The paired anterior ends of the pelvic girdle *(ilium)* are united with the enlarged transverse process of the ninth or sacral vertebra. Other bones of this girdle, arising from cartilage, are the *ischium* and *pubis*. The pubis alone does not ossify. The ilium forms a cup, the *acetabulum*, which receives the head of the *femur* of the hindlimb. The bones of both anterior and posterior girdles and limbs arise by the ossification of preformed cartilage (i.e., endochondral in origin). There is evidence that appendicular muscles of some amphibia may develop in situ from somatic mesoderm.

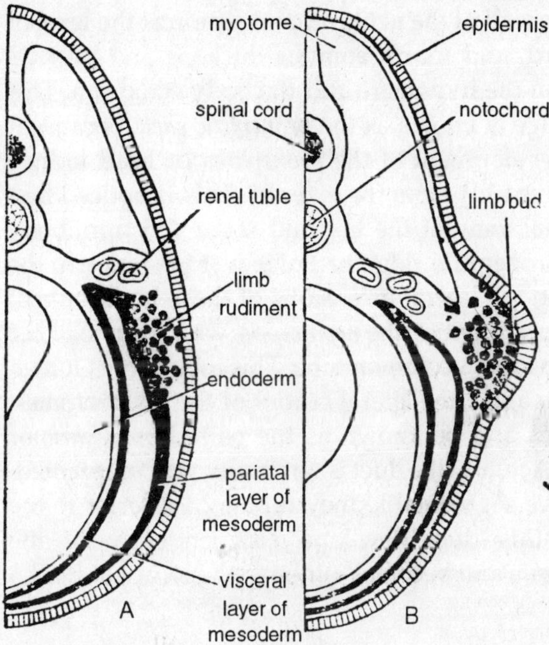

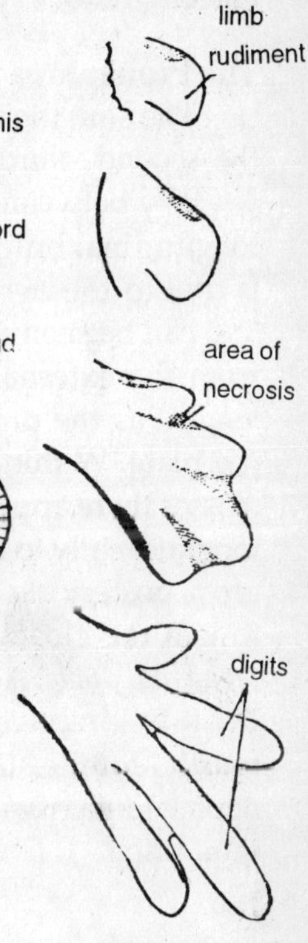

Fig. 6.5. Origin of limb-mesoderm and formation of a limb-bud of an amphib an embryo.

Fig. 6.6. Differentiation of a limb-bud into hind-limb of a vertebrate embryo.

The Mesomere (Intermediate Cell Mass)

Early in embryonic development the upper level of the lateral plate mesoderm becomes separated off from the ventral sheets of mesoderm, to give rise to the excretory and parts of the reproductive systems. This is known as the *intermediate cell mass* or *mesomere* because of its position relative to the other masses of mesoderm. It is also known as the nephrotome because it gives rise to excretory units, both the larval pronephros and the functional mesonephros of the frog. The derivatives of this mesoderm will now be described; the description of the development of the hypomere and its derivatives, the coelom and circulatory system, will be deferred until later in the book.

The Pronephros or Head Kidney

The somatic wall of the nephrotomal region at the level of the second, third, and fourth somites thickens and projects laterally between the hypomere and the body ectoderm. This longitudinal bulge is known as the *pronephric shelf,* because it is due to the development of the *pronephros* or head kidney and can be seen readily from the exterior. This is noticed first from the external view at the tail bud stage (2.5 mm. body length) as the pronephric ridge or bulge just posterior to the gill plate. Within the lateral extensions of these nephrotomal masses there appear cavities, the *nephrocoels,* which run together longitudinally to form a common tube. This tube and its lumen grow posteriorly along the lateral border of the nephrotomes, to join the cloaca and be known as the paired *pronephric* or *segmental ducts.* Actually the duct is not in any way segmented, but it will receive segmental kidney tubules. Posterior to the fourth somite this duct develops independently of the nephrotomal tissue and becomes joined to the cloaca at the 4.5 mm. stage.

Three pairs of *pronephric tubules,* one at the level of each of the somites II, III, IV, appear between the segmental duct and the coelom. The original connection of the nephrotomes with the somites is lost in favour of a nephrogenic cord. Finally each tubule acquires an opening, the *nephrostome,* into the adjacent coelom. This opening shortly acquires cilia and becomes funnel-shaped, by the time it teaches the 5 mm. body length

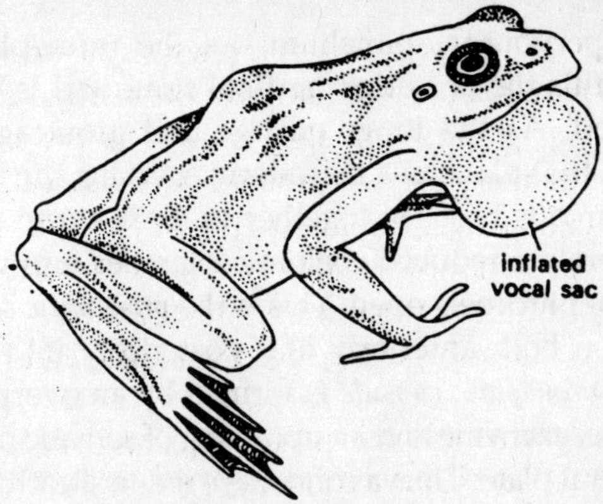

Fig. 6.7. A to G stages of the development of a mesonephric renal unit (uriniferous tubule) of frog. H is a fully formed mesonephric tubule of frog.

stage. The pronephric tubules lengthen and become coiled (convoluted) and at the 6 mm. stage are embedded in the mesenchyme and sinuses of the developing posterior cardinal vein. The entire pronephric mass is surrounded by the connective tissue of the pronephric capsule, derived from the adjacent somatic mesoderm and myotome.

Between each nephrotome and the dorsal aorta, suspended into the body cavity and surrounded by splanchnic mesoderm, a very rudimentary capillary network known as the *glomus* arises from the dorsal aorta. This appears to be an abortive attempt to develop a glomerulus characteristic of the adult mesonephric kidney, and occurs in response to inductive influences from the non-functioning pronephric tubules.

The *pronephros* is structurally much like the functional kidney of the frog. except that it is much less complex. It acquires a rich vascular supply, both arterial and venous, but has no outlet. It is doubtful that it ever functions as an excretory organ but it attains its maximum development at about the 12 mm. stage. Along with the glomus the elongated and coiled tubes of the pronephros constitute a formidable mass of tissue which projects into the dorsal body cavity and is surrounded largely by peritoneal epithelium. As the pronephric mass expands, it fills the posterior cardinal sinus and is bathed in venous blood. As the lungs enlarge and grow against the pronephros, the splanchnic mesoderm covering the lungs and the pronephros is brought together to form a trap or pocket which is merely a reduced coelomic chamber into which the temporary nephrocoels open. This is the *pronephric chamber*. It remains open both anteriorly and posteriorly into the lung chamber. *A pronephric capsule* is formed by an overgrowth of myotomic mesenchyme and an upfolding of somatic mesoderm from the lateral plate. Thus a connective tissue sheath is formed around the embryonic head kidney.

By the 11 mm body length stage the pronephroi are large and conspicuous bodies with blind tubular outgrowths; but by the 20 mm stage they have begun to degenerate and neither the tubules at this level nor the related glomi remain. The mesonephric kidney develops rapidly and takes over the increasingly important excretory functions.

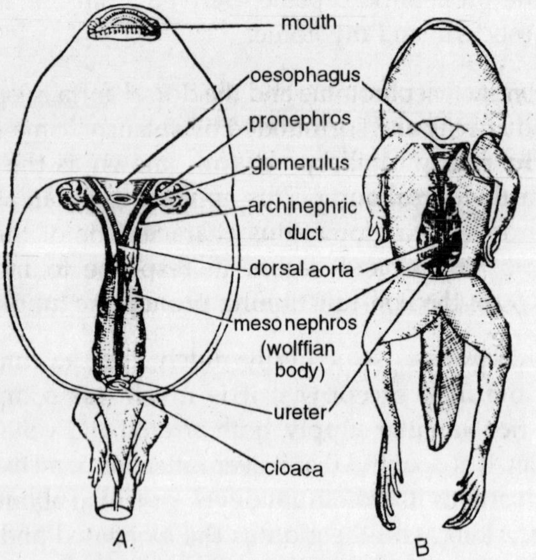

Fig. 6.8. Replacement of pronephros of tadpole (A) by mesonephros of tailed frog (B) during metamorphosis.

The Mesonephros or Wolffian Body

The nephrotomal mass posterior to the pronephros and mesial to the segmental duct gives rise to the *mesonephros* or *Wolffian body* which is the functional kidney of the adult frog. It extends from the level of somites VII through XII and begins to develop at the 8 to 10 mm. stage.

In this region, as in the more anterior regions, there are segmental nephrotomes which merge to form a pair of continuous *nephrotomal* masses from the seventh to the twelfth somites. The mesonephros is therefore of both somatic and splanchnic mesodermal origin. At the level of each of these somites there develop several *nephrotomes*, each with a separate *nephrocoel*. The nephrocoels at this level are known as the *mesonephric vesicles* which become convoluted and constricted so as to give rise to primary, secondary, and tertiary units, in that consecutive order. It is obvious, therefore, that the kidney units of the mesonephros are more complicated than those of the pronephros, from the very beginning of their formation. They are not actually metameric.

The *primary mesonephric visicle* gives rise to a dorsal evagination which becomes the *secondary* unit of the *mesonephric vesicle*. Following this there is another evagination from the ventro-lateral side of the primary vesicle, toward the segmental duct with which it becomes continuous. This is known as the *inner* tubule of the *tertiary mesonephric vesicle*. It becomes greatly coiled and encroaches upon the developing and adjacent cardinal vein. The inner tubules become the tubular portion of the adult kidney. The *segmental duct* from this level posteriorly is known as the *Wolffian* or *mesonephric duct*, and this will remain as the excretory duct or ureter.

A third evagination, the *outer tubule*, develops ventrally from the mesonephric unit. The cavity of this tubule joins the coelom by way of the nephrostome. However, by the 20 mm. stage this unit has broken away from the rest of the mesonephric unit and has changed its connection from the mesonephros to the lateral division of the *cardinal vein.* This vessel becomes the *renal portal vein.* This connection can be demonstrated very easily in the recently metamorphosed adult frog where persistent ciliated nephrostomes can be seen on the ventral face of the kidneys and which can be shown to maintain direct connections with venous capillaries of the kidneys. The outer tubule aids in forming *Bowman's capsule* around each golmerulus. The ciliated nephrostome then arises along with the mesonephric unit, develops a normal coelomic connection, and then shifts its relationship to become associated with blood sinuses of the *posterior cardinal vein.* This occurs within the kidney, at about the 15 mm. stage. The original nephrostomes remain in the adult frog on the ventral face of the kidney as ciliated *peritoneal funnels* which are readily seen. These convey coelomic fluids directly into the blood sinuses of the kidney. They might still be regarded as accessory excretory structures. More nephrostomes appear than can be accounted for in the above manner and it is not known whether the extra ones arise by splitting of the original ones or by additional peritoneal invaginations to the mesonephric tubules.

The mesonephric tubules which remain to give rise to the *uriniferous tubules* of the functional and adult mesonephric kidney are elongated considerably. They form spherical masses around developing blood capillaries emanating from both the renal artery (from the dorsal aorta) and from the renal portal vein. These form capillary networks which are both arterial and venous and constitute true *glomeruli,* having no connection with the coelom as do the glomi of the pronephric level. Each glomerulus is surrounded by the thin-walled *Bowman's capsule* of the fully formed *Malpighian body* (renal corpuscle) within the mesonephric kidney. By the time of metamorphosis this mesonephric kidney is fully formed and functional, ready to assume the increased excretory load of a terrestrial organism.

The Adrenal Glands

Most of the endocrine glands are derived from the pharyngeal endoderm or brain ectoderm, but the *adrenals* are derived from both ectoderm and mesoderm. They are a complex of two endocrine glands and are not related functionally to the excretory system but can be described conveniently at this time because of their proximity of origin and their final position in the adult. In the adult frog this composite gland appears as a thin layer of yellowish brown granular material on the ventral face and closely adherent to each kidney. It is composed of both a medullary (suprarenal) substance and a cortical (interrenal) substance, so called because of their positional relationship in the higher forms rather than in the frog.

The *Adrenal Cortex (Interrenal Gland)*. The cortical cells of the adrenal gland may be seen as early as the 10 mm. body length stage, and probably arise from the coelomic mesothelium. They are aggregated into round or elliptical bodies lying on the dorsal surface of the coelomic cavity, just ventral to the dorsal aorta. The original mass measures about 0.8 mm. in length. It contains a number of nuclei embedded in a syncytial mass of acidophilic tissue. The contained granules may be pink, yellow, or black. By the 12 mm stage paired gonad primordia may be seen, also in the dorsal coelomic epithelium but in a more lateral position. By the 13 mm stage the cortical masses have doubled. They form a cap around the base of the upper four-fifths on either side of the dorsal aorta, just posterior to the site of formation of a single aorta. At this stage the cortical material may be seen intermingled with, or in close proximity to, the gonad anlage.

By the 37 mm body length stage the adrenal cortex consists of lobed masses of cells surrounded by the notochord, the body wall, and the Wolffian duct below. Groups of cortical cells are, by this time, encapsulated and come into intimate association with the mesonephros. By metamorphosis these groups of cortical cells are interspersed with strands of medullary cords. The groups migrate from the dorso-mesial side of the mesonephros to its ventral surface where they are found in the adult, extending almost the full length of the kidney. It now consists of clear-cut areas of medullary tissue associated with circumscribed lobes of cortical tissue. This cortical tissue characteristically contains distinct Stilling cells, never to be found in the medullary tissue. We must conclude, therefore, that since the frog adrenal contains both cortical and medullary material its origin is from both mesoderm and ectoderm.

The Reproductive System

The Gonoducts : The Male. In the male the *mesonephric* or *Wolffian duct* acts as both an excretory organ (i.e., ureter) and a reproductive duct (i.e., *vas deferens*) since all spermatozoa pass from the testis into the *uriniferous tubule* of the mesonephric kidney by way of the *vasa efferentia*. These tubules are modified anterior mesonephric tubules which convey spermatozoa into the *Malpighian corpuscle* by way of permanent opening into the *Bowman's capsule*. The sperm then continue to the cloaca by way of the uriniferous tubule and the mesonephric duct. There is therefore no separate *vas deferens*, this mesonephric or Wolffian duct acting in the normal capacity of a vas deferens. In the male frog, then, it is a true *urogenital* duct from the beginning.

As the mesonephric (Wolffian duct or vas deferens) duct reaches the cloaca it enlarges and becomes somewhat glandular, and is known as the *seminal vesicle*. It is within this vesicle that the sperm masses are retained during amplexus and from which they are ejected into the water during oviposition.

The *Müllerian duct*, which is the male homologue of the oviduct, develops late in embryogenesis from a longitudinal ridge of cells which project into the coelomic cavity just ventral and lateral to the pronephric region and the Wolffian duct. This ridge sometimes develops a tube and extends from the cloaca to a point near the junction of the lungs, anterior liver lobes and heart. It is covered with and suspended by a thin layer of peritoneal epithelium. Originally this Mullerian duct was described as originating from a longitudinal splitting of the segmental duct, but this has been doubted recently. The duct often degenerates in the male adult but there may be vestiges in higher vertebrates such as the appendix, testis, and prostatic utricle. Its homology to the female oviduct can be demonstrated by treating the male with estrogens which cause it to hypertrophy, and to have the appearance of an oviduct.

The Female. The *mesonephric* or *Wolffian duct* of the female functions exclusively as a *ureter*. The *oviducts* arise is a manner similar to the Mullerian ducts of the male but they acquire a lumen which is surrounded by ciliated and glandular epithelium and muscular walls. It is suspended to the dorsal body wall by a double fold of peritoneum. The anterior end of each oviduct consists of a persistent group of nephrostomes which become the fimbriated and highly elastic *infundibulum* or *ostium tuba*. The oviduct shows slight differential gland development and convolutions from the anterior to the posterior, where it becomes the very distensible *uterus*. The two uteri open separately into the cloaca.

The body cavity of the female develops an abundant supply of cilia, in response to the appearance of an ovarian hormone. These cilia may therefore be regarded as secondary sexual characters of the female, appearing shortly after the time of metamorphosis.

The Gonads. The paired gonads arise as a single median sex cell ridge dorsal to the midgut at about the time of hatching (6 mm stage). These cells presumably arise from the *yolk sac splanchnopleure*, or from the gut endoderm. Some evidence in favor of this theory is the fact that the primordial germ cells are heavily laden with yolk. As the two lateral mesodermal plates converge above the gut, this group of cells migrates to a position directly dorsal to the dorsal mesentery, between the posterior cardinal veins at the 8 to 9 mm stage. Since this ridge of primitive gonadal material is rather long, it is now called the *sex cell cord*. This shortly divides longitudinally into two, and each part moves ventro-laterally to a position between the developing Wolffian body and the dorsal mesentery projecting into the body cavity as the *genital ridges*. Each mass will enlarge and project ventrally into the coelomic cavity, carrying with it a covering of peritoneal epithelium. This membrane becomes double and suspends the gonad to the body wall as the gonad to the body wall as the gonad grows into the coelom. This double membrane is known as the *mesorchium* in the male and the mesovarium in the female.

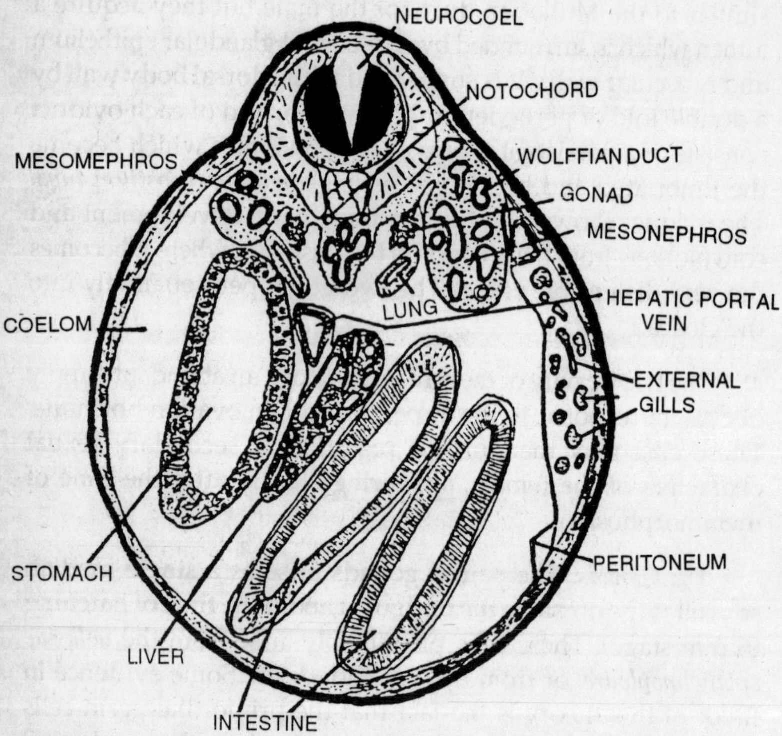

Fig. 6.9. Schematized diagram through the level of the gonad primordium of the 11 mm frog tadpole.

These coelomic-projections of pre-gonad material are known as *genital ridges.* Sex is, of course, not distinguishable as the gonad primordium shows no particular cellular differentiation at this time. The general organization of both presumptive ovary and testis is one of an enlarged sac with *germ cell primordia* around the periphery and the *primary genital cavity* in the center.

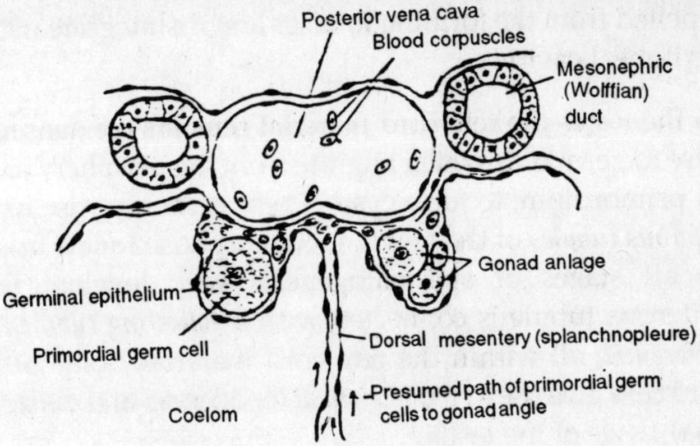

Fig. 6.10. Gonad primordia of the 11 mm frog tadpole.

The *rete cords* now appear as strands of cells which arise from the developing Malpighian bodies of the mesonephros and grow into the primary genital cavity of the undifferentiated gonad. These strands unfortunately are referred to in some textbooks as sexual cords or sex cords, but these terms are confused too easily with sex cell ridge or sex cell cords, terms just used. They properly are called *rete cords.*

Sex differentiation occurs before the time of metamorphosis, at about the 30 mm body length stage. The gonads show macroscopic differences at this time. In the case of the potential ovary the cells around the periphery of the gonad primordium multiply and the rete cord strands, which have invaded the genital cavity, begin to develop their own cavities, which are known as the *secondary genital* cavities. These become confluent and form from seven to twelve large ovarian sacs, each connected with the others. The cellular material of the rete cords then gives rise to clusters of cells, one of which begins to grow toward the primary oocyte stage. The others (primordial germ cells) become the surrounding nurse or follicle cells. All these cells probably are derived from the outer germinal epithelium. They are found at all times as clusters of potential ovarian follicles, both nurse cells and ova. Many primordial germ cells are expelled from the forming follicles and disintegrate within the peritoneal cavity.

In the testes the rete cord material remains condensed to give rise to germ cells which migrate from the periphery to the gonad primordium, to form cysts. These cysts give rise to the *seminiferous tubules* of the newly metamorphosed male frog in which all stages of spermatogenesis will develop. Each seminiferous tubule is connected with a *collecting tubule* and *vasa efferentia*, all within the rete cord material. Some of the rete cord cells give rise to the *interstitial* (endocrine) and *connective* (stroma) *tissue* of the testis.

Fat bodies develop from one-third to one-half of the anterior ends of either of the genital ridges. These are storage masses built up during the summer feeding periods in anticipation of hibernation and the subsequent active spring breeding period. In closely related toads of certain species the anterior end of the testis often may develop a rudimentary ovary, with oocytes, which will respond to hormonal treatment much in the manner of a normal ovary. This is known as *Bidder's organ*. Isolated oocytes have been found in the seminiferous tubules of the otherwise normal male frog, and hermaphrodites have been described. These facts simply emphasize the fundamental similarity of the frog ovary and testis, both in origin and in early development.

THE HYPOMERE (LATERAL PLATE MESODERM)

The Coelom (Splanchnocoel)

This *splanchnocoelic* cavity arises as irregular spaces in the lateral plate mesoderm, ventral to the mesomere. These spaces become confluent to form a continuous split, the *coelom*. The cavity therefore develops between an outer *somatic* and an inner *splanchnic* layer of mesoderm. The earliest appearance of any coelomic space is in the tail bud stage, and in the anterior heart mesenchymal area. There shortly develop three continuous divisions of the coelom, the muscle coelom (*myocoel* of the somite), kidney coelom (*nephrocoel* of the intermediate cell mass), and gut coelom (*splanchnocoel* of the lateral plate). In addition, there is the heart coelom or *pericardial* cavity.

By the time of hatching, the coelomic space (splanchnocoel) on one side merges ventrally with that on the other side (except at the extremities of the embryo) to form a horseshoe-shaped cavity surrounding the yolk and enteron. The only region where the coelom does not become continuous is the area dorsal to the enteron (gut) or ventral to the notochord. At this region the splanchnic mesoderm comes together to form the *dorsal mesentery.* This double sheet of splanchnic mesoderm then acts as a suspensory mesentery for the gut, and it continues around to enclose the enteron and the yolk. This lining of the visceral cavity and covering of the gut will become the *peritoneal epithelium.*

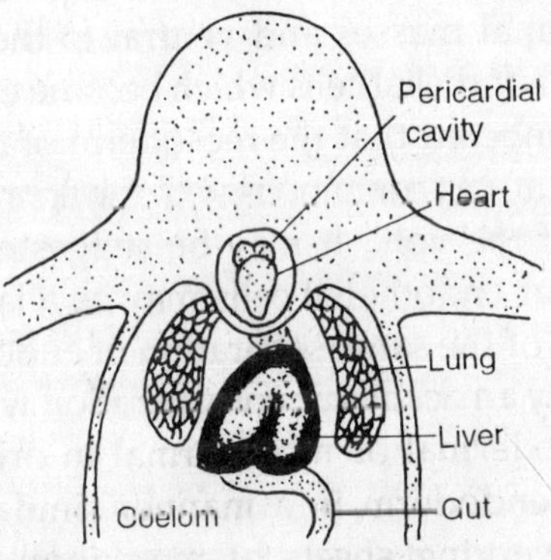

Fig. 6.11. The coelom and its derivatives in the frog.

Just posterior to the pharynx the coelomic space extends into the heart mesenchyme as the pericardial cavity, to be described below. During the tail bud stage, for instance, one could inject a coloured dye into the coelom at the yolk level and this would be carried anteriorly to fill the forming pericardial cavity. The frog develops lung respiration by the time of metamorphosis but it does not develop a diaphragm to separate the coelom further into a pleural and peritoneal cavity as in the case of higher vertebrates. These cavities remain continuous, in the frog, as the *pleuro-peritoneal cavity*, but the pericardial cavity becomes cut off entirely from the embryonic coelom. This is accomplished by the aid of the developing ductus *Cuvieri* which helps to form the *septum transversum* at about the time of metamorphosis.

The Heart

The heart rudiment develops early from splanchnic mesenchymal cells which grow ventrally toward the mid-line below the pharynx. There they aggregate to form paired spaces (anterior coelomic spaces) at about the tail bud stage. This seems to occur under the inductive influence of the archenteric floor. In the meantime there appear scattered and possibly endodermal cells dorsal to the point of junction of the converging mesodermal masses and ventral to the point of junction of the converging mesodermal masses and ventral to the pharynx. These are known as endothelial cells which become endocardial cells. It will be remembered that the mesoderm of the ventral lip of the blastopore is at first continuous with the floor endoderm of the gut by way of the yolk. It may be suggested that the separation of the heart endothelial cells may be a later (hence more anterior) phase of the same separation of endoderm and mesoderm. It is purely an academic consideration whether the endocardium is endodermal or mesodermal in origin, for it arises from original endoderm in a manner similar to most mesoderm. The converging sheets of mesoderm then fuse above and below, enclosing these endothelial cells. The wall of the inner tube thus formed becomes greatly thickened during this convergence and is known as the *myocardium*. It will begin to give rise to the striated, syncytial, and involuntary muscles of the heart by the 5 mm body length stage of development. The enclosed endothelial cells will form the lining of the heart known as the *endocardium*. The cavity surrounding the myocardium is the *pericardial* cavity which itself is enclosed by an outer (splanchnic) *pericardium*. The pericardial portion of the embryonic coelom is separated from the peritoneal cavity by growth of the common cardinal veins, the liver, and peritoneal folds, all of which comprise the septum *transversum*. This process is completed during metamorphosis.

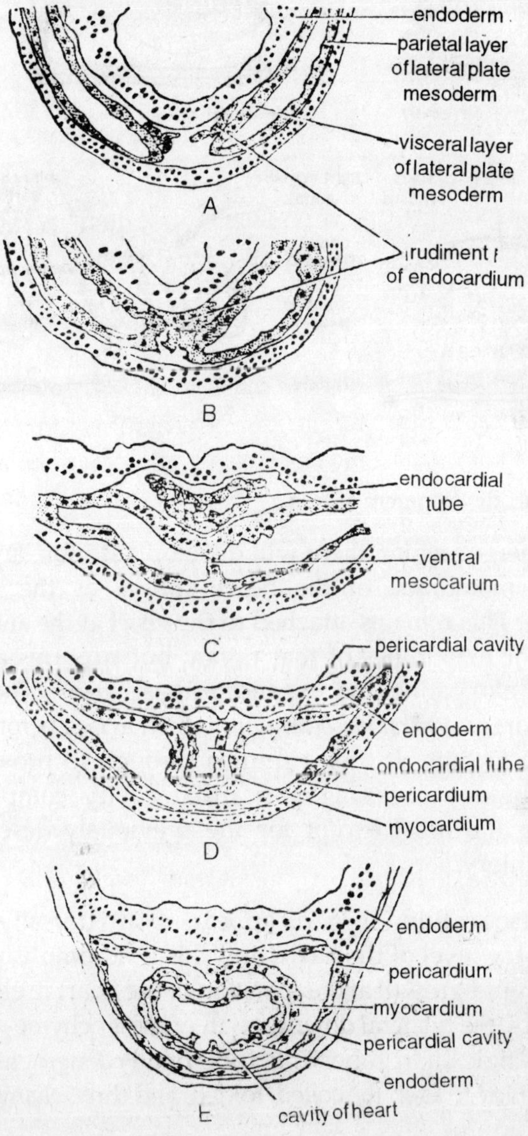

Fig. 6.12. Development of the heart in amphibian embryos.

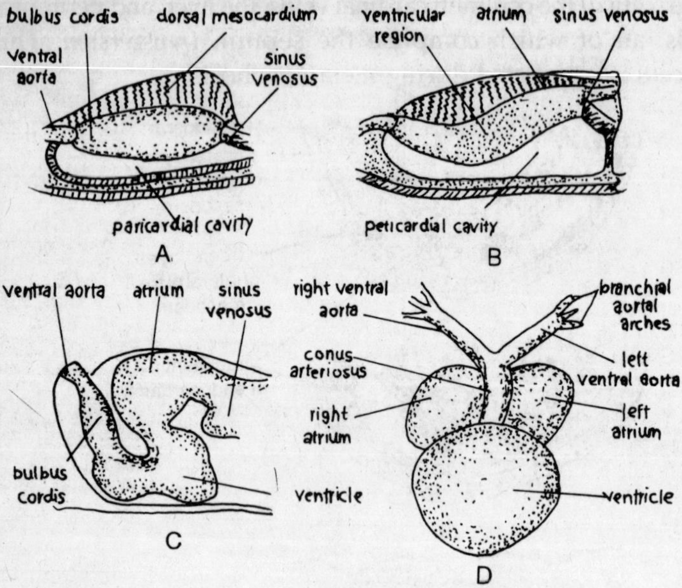

Fig. 6.13. Heart development of frog.

As in the gut region there will develop a double-layered suspensory membrane of the heart known as the *dorsal mesocardium*. This remains attached to the heart at the anterior and posterior extremities of that organ, but ruptures at the level of the elongating and expanding heart itself. There is also a temporary *ventral mesocardium*, which arises before the dorsal mesocardium. It soon ruptures through to provide a single continuous enclosing pericardial cavity completely surrounding the heart except for the previously described dorsal mesentery.

The mesocardium then disappears both dorsally and ventrally at the level of the heart, and this is accomplished in part by the rapid growth and expansion of the heart itself. The heart arises by the bilateral downgrowth of mesenchyme which becomes a single short tube. This tube finally lengthens and becomes divided to form the coiled, folded, and three-chambered heart of the adult.

The early differentiated heart consists of a simple tube which becomes divided into three chambers. The fused pair of vitelline veins becomes the *meatus venosus* which empties into the first division of the heart, the sinus venosus. Then follows (more anteriorly) the thin-walled *atrium*, the thick-walled ventricle, and finally the *bulbus arteriosus* (bulbus aortae) which leads to the *ventral aorta* or *truncus arteriosus*.

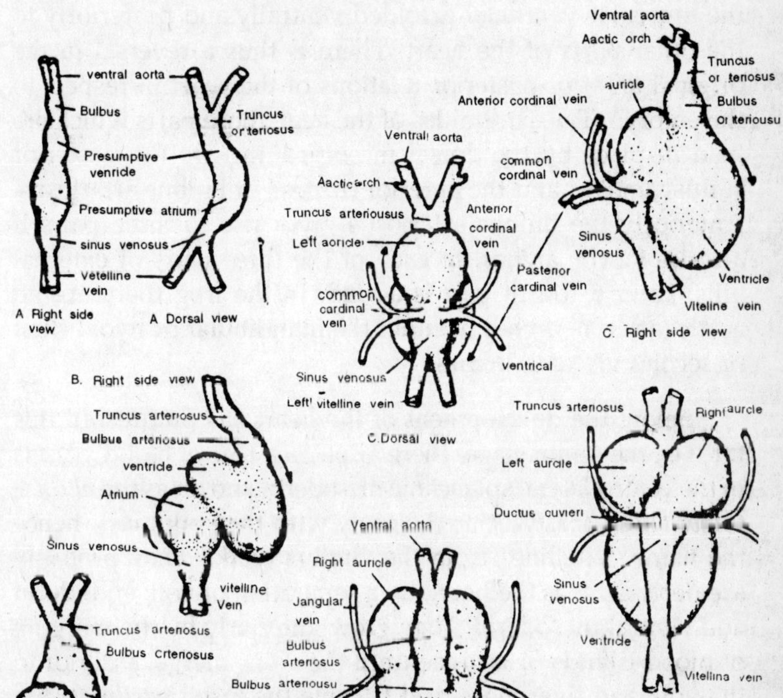

Fig. 6.14. Development of the amphibian heart.

The development of the heart is accomplished within the limited space provided by the surrounding body tissues and organs, so that this organ becomes a reversed "S" coiled tube. With the embryo facing to the right, and looking at it from the right side, the shape of the heart would be something like this: (the figure on last page)

The posterior part of the original tube becomes folded dorsally and anteriorly to the more anterior part of the original tube. Conversely, the anterior part (which is destined to become the muscular ventricle) is folded ventrally and posteriorly to the other parts of the heart. There is thus a reversal in the original anterior-posterior relations of the heart in respect to the embryo. The extremities of the heat, those parts which are held in place by the dorsal mesocardium, are the posterior sunus venosus and the anterior truncus or bulbus arteriosus. Anteriorly the bulbus arteriosus gives rise to three pairs if arteries (aortic arches) to each of the three pairs of external gills. Later a fourth pair is added. In the frog there are no aortic (arterial) arches in either the mandibular or hyoid (first or second visceral) arches.

Before the development of the heart has progressed this far (i.e., the 4 mm stage) there appears a pair of blood vessels in the ventro-lateral splanchnic mesoderm known as the *vitelline veins*. These are associated closely with the yolk mass, hence the name "vitelline" from the vitellus (yolk). They probably are derived, as is the heart, by a separation of yolk endoderm and ventral mesoderm. They grow anteriorly by the merging of blood islands or lacunae until the veins merge, anterior to the yolk and liver mass, and become the fused *vitelline veins*. This point of fusion of the vitelline veins is called the *sinus venosus* from the very beginning. It represents the most posterior limit of the heart. The part of the heart into which the sinus venosus empties is the *atrium* or the undivided embryonic auricles. The atrium and subsequent auricles remain thin-walled and become very elastic. Shortly (i.e., at the 7 mm body length stage) there grows down from the roof of the atrial chamber a longitudinal sheet of endocardium which is known as the *interatrial septum*. This divides the atrium into a *right* and *left* auricle, and the sinus venosus now leads into the right auricle. Eventually the left auricle will receive blood from the pulmonary veins.

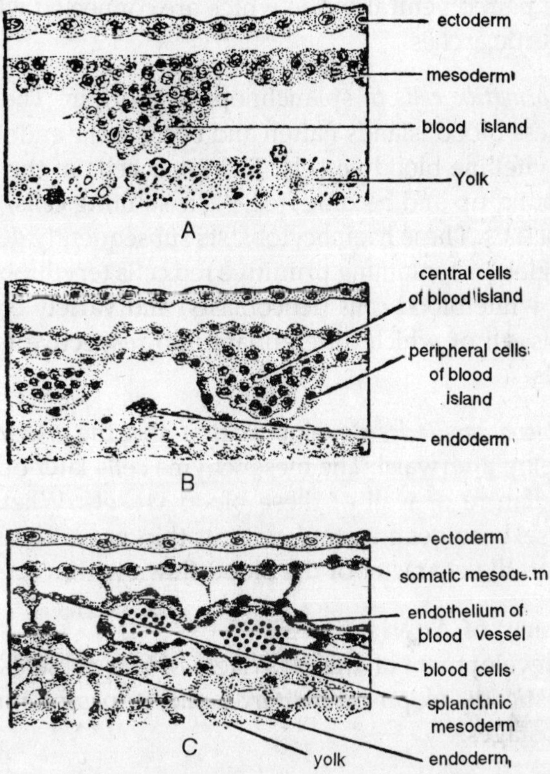

Fig. 6.15. Differentiation of blood islands and formation of blood cells and endothelium of blood vessels of a vertebrate embryo.

The *ventricle* or anterior portion of the original heart tube becomes very thick-walled and muscular and is never divided longitudinally into two as it is among mammals. There is a transverse division groove between the atrium and the ventricle at the folds in the inverted "S" coil just described. The ventricle is therefore folded caudally and ventrally to the atrium and leads into the truncus arteriosus which is suspended to the dorsal body wall by the permanent *dorsal mesocardium* at this level. The truncus arteriosus becomes partially subdivided by a vertical and longitudinal septum (the spiral valve) and gives rise to the paired ventral aortae which are connected with the various aortic arches.

Vasoformative cells of splanchnopleuric origin. The outer cells of these blood islands flatten and become the endothelial lining of vitelline blood vessels. The inner cells of the blood islands round up and become free, basic staining cells, called *haemocytoblasts*. These haemocytoblasts subsequently develop in haemogloblin containing primitive red cells (erythroblasts); primitive white blood cells (leucoblasts) and variety of other blood cells, all of which differentiate into respective mature blood cells.

The *blood capillaries* develop at first and the *arteries* and *veins* develop afterward. The mesenchyme cells later on form the musculature of wall of these blood vessels. Whether, a blood vessel becomes an artery, with its thicker wall or a vein, depends on the character of the blood flows within it.

Development of Arterial System

The development of arterial system of frog presents quite characteristic development patterns and it passes through following stages:

(1) The arterial system during external-gills stage of tadpole : While tadpole is breathing by means of gills, its circulation resembles in all respects with that of fish. The *venous blood* returned from the body enters the posterior end of the heart or sinus venosus; from this it passes into the second or auricular chamber, thence to the ventricle, and from that to the truncus arteriosus. The truncus arteriosus on reaching the anterior end of the pericardial cavity, bifurcates into right and left branches. Each of these again divide into *first*, *second* and *third*

afferent branchial arteries, which carry venous blood into the external gills and their branches. The *fourth afferent branchial artery* remains absent.

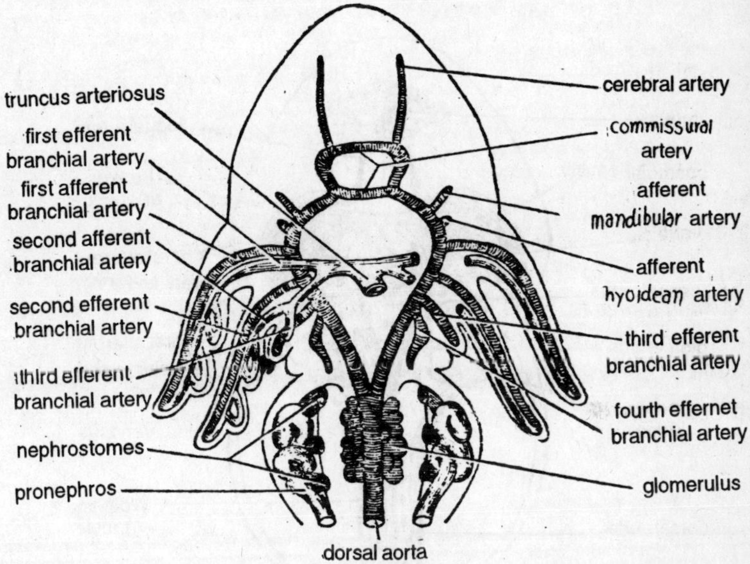

Fig. 6.16. The arrangement of afferent and efferent branchial arteries in a tadpole larva in its external gill stage.

In the external gills, the venous blood passes through short wide capillary loops into four pairs of *efferent branchial arteries,* which thus, form along with afferent branchial arteries, three pairs of *complete branchial arches* and one pair of incomplete (i.e., *fourth) branchial arches.* All efferent branchial arteries open into a pair of *dorsal aorta* in the roof of pharynx. The dorsal aortae of the two sides run forward as the *carotid arteries* to supply the head and brain, and also run backwards in the roof of the pharynx, the aortae of the two sides meeting and uniting about the junction of head and body to form the single *systemic arch* which supplies arterial blood to all body organs.

In external gill stage, there are thus six sets of branchial vessels on each side of the pharynx of tadpole—of these, three in the first, second and third branchial arches are incomplete and two, in the hyoid and mandibular arches (i.e., *mandibular* and *hyoidean branchial arches)* are rudimentary.

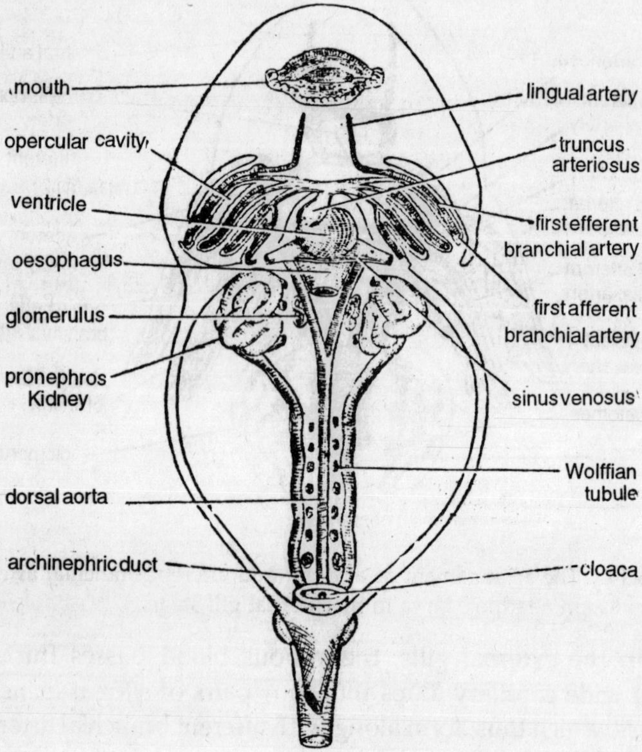

Fig. 6.17. The branchial arches of a tadpole in internal gill stage. The development of pronephric and mesonephric kidneys is also illustrated in the diagram.

(2) The arterial system during internal-gills stage of tadpole : After the external gills are shriveled up, and opercular or internal gills become well-established in the tadpole following modifications occur in the blood circulation of tadpole—The truncus arteriosus divides at once into right and left branches, which run straight outward in the floor of the pharynx. Each of these branches divides, after a short course into three vessels, and the hind most vessel again into two. In this way the four

afferent branchial vessels of the first, second, third and fourth branchial arches are formed respectively.

Each afferent vessel runs outwards and upwards in its own arch. The efferent branchial vessels lie immediately in front of the corresponding afferent vessels with which they are connected by very numerous capillary loops in the substance of the internal gills. At their upper ends the efferent vessels open, as before into the dorsal aorta.

As the lungs become considerably developed during late tadpole stage, they receive blood by the *pulmonary arteries* which are formed by branches of fourth efferent branchial vessels. The blood from the lungs returns directly to heard by a pair of pulmonary veins which into left auricle.

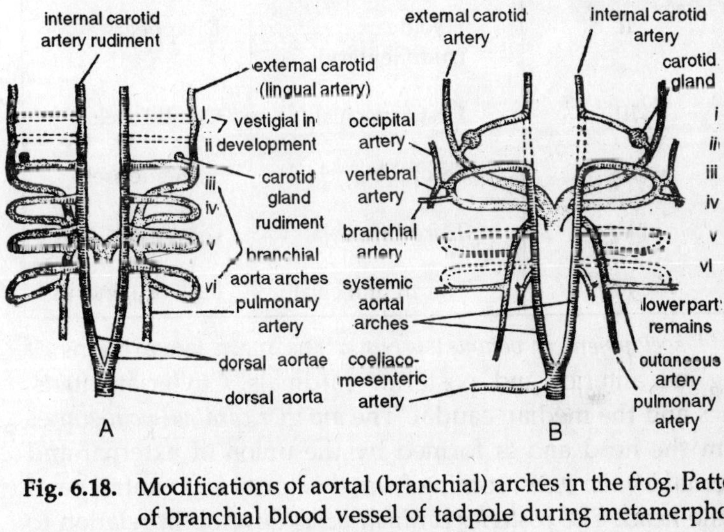

Fig. 6.18. Modifications of aortal (branchial) arches in the frog. Pattern of branchial blood vessel of tadpole during metamerphosis (A) and after metamorphosis (B).

(3) The changes in arterial system during metamorphosis : With the atrophy of tadpole gills during metamorphosis, considerable changes occur in arterial system. Of the four aortic arches, the *first branchial arch* persists as the *carotid arch* of the adult frog; the *lingual artery* is a branch from the ventral end of the efferent

vessel of the arch, and is present from an early stage of development.

The *second branchial arch* becomes *systemic* arch of the frog. Its dorsal end remains connected with the carotid arch. The *third aortic arch* loses its connection with the aorta and finally disappears altogether. The *fourth aortic arch* also loses its connection with the aorta, but persists as *pulmocutaneous* arch of the adult, from which both *pulmonary* and *cutaneous arteries* arise. All of these changes can be summarized as follows:

Aortic arch	*Embryonic (Tadpole)*	*Adult (Frog)*
I	Mandibular (rudimentary)	Disappears
II	Hyoidean (rudimentary)	Disappears
III	First branchial	Carotid arch
IV	Second branchial	Systemic arch
V	Third branchial	Disappears
VI	Fourth branchial	Pulmocutaneous

Development of venous system : The main larval veins of frog are anterior and posterior cardinals, Cuvierian ducts, iliacs and the median caudal. The *anterior cardinal vein* comes from the head and is formed by the union of external and internal *jugular veins,* coming from the dorsal and ventral parts of the head. The *posterior cardinal veins* develop in relation to the pronephros and collect the blood from the posterior body organs. The anterior and posterior cardinal of each side unite to form the *common cardinal vein,* called *ductus Cuvieri.* The *lateral abdominal veins* extend from the posterior end to open into the ductus Cuvieri. The *iliac veins* from the hind limbs open into them. The *caudal vein* comes from tail and join left and right posterior cardinal veins.

Change in venous system during metamorphosis : During metamorphosis, following changes occur in lateral venous system:

(i) With the appearance of liver, some of the vitelline veins become degenerated and remaining ones develop into *hepatic portal system.*

(ii) With the replacement of pronephros by mesonephros, the right and left posterior cardinals unite to form a *median cardinal vein* which collect blood from mesonephros and become *posterior vena cava,* which open into sinus venous. Meanwhile, at the outer lateral borders of both mesonephros; arise new veins which become connected with median cardinal vein by many capillaries which run through the substances of kidneys and are called *renal veins.* The lateral veins are called *renal portal veins.*

(iii) With the appearance of legs, each renal portal veins is joined by a *iliac vein,* which divides at its distal end into the *femoral* and *sciatic veins.*

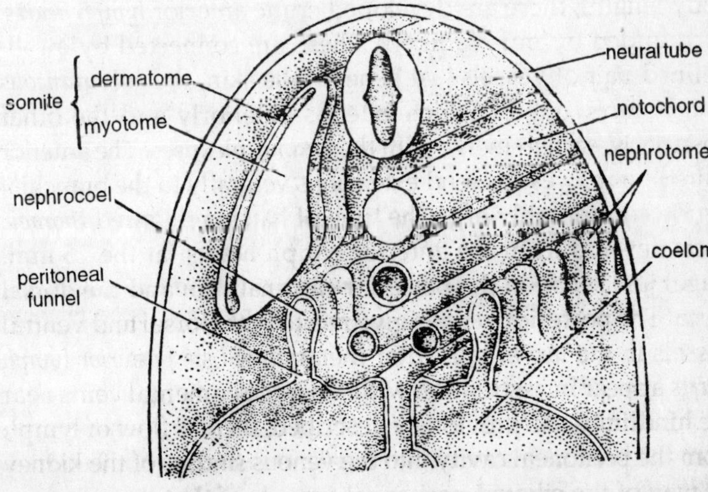

Fig. 6.19. Stereogram of vertebrate embryo showing the differentiation of nephrotomes.

(iv) When tail is lost, the part of the posterior vena cava

posterior to the kidneys (mesonephroi) vanishes, so that most of the blood from posterior body region has to pass through the renal portal vessels and the abdominal veins.

The Lymphatic System. The lymphatic system of the frog is made up of very large *lymph spaces* in various parts of the body, without well-defined connecting lymph vessels such as occur in mammals. However, all of the four lymph hearts are protected by valves so that the lymph always passes from a lymph heart to a lymph or blood vessel, and never in the reverse direction. There are no muscular coats to the lymph spaces, but they are lined with flat endothelial cells and connective tissue. The connective tissue forms septa which divide the lymph spaces and attach them to the underlying muscle.

Between the third and fourth somites between the peritoneum and the integument, at the time of hatching (6 mm body length), there are developed crude anterior *lymph hearts,* surrounded by muscle fibers. These are connected by an ill-defined pair of vessels just beneath the skin, the *subcutaneous lymph spaces,* one of which extends anteriorly and the other posteriorly, each to merge with the venous plexuses. The anterior paired lymph vessels send extensions ventrally to the branchial region considerably after the time of hatching. Paired *thoracic ducts* develop from the anterior lymph hearts (at the 26 mm. stage) just between the posterior cardinal vein and the dorsal aorta. The posterior vessels give rise to both dorsal and ventral vessels in the larval tail. After metamorphosis *posterior lymph hearts* appear in association with the intersegmental veins near the hindlimbs. As previously described, there is flow of lymph from the peritoneal cavity into the venous sinuses of the kidney by way of the ciliated peritoneal funnels of that organ.

The Spleen

This hematopoietic organ arises in the vicinity of the stomach at about the 10 mm body length stage as a cluster of cells within the dorsal mesentery. By the 15 mm stage the spleen is seen as a definite projection from the mesentery, covered by coelomic epithelium. Apparently some wandering cells from the intestinal epithelium invade the spleen at this time. By the 30 mm stage these cells become reticular and are vascularized and can be recognized as splenic cells, and the organ as the ovoid spleen of the adult.

7

The Endodermal Derivatives

In most of vertebrates endoderm is formed from the micromeres and endoderm from the megameres. But during gastulation as the development proceeds further these cells are shifted from surface layer to their final destinations, wegameres form the inner lining of archenteron and are transformed into the endoderm.

The endoderm gives rise to all structures associated with the original archenteron, from the mouth *(ectodermal stomodeum)* to the anus *(ectodermal proctodeum)*, and all of its derivatives. It must be emphasized, however, that the endoderm contributes only the lining epithlium of these structures and that ectoderm and mesoderm may also be involved. It is nevertheless convenient to describe in sequence from the anterior to the posterior (foregut, midgut and hindgut) those structures whose linings are endodermal in origin.

The Mouth

Mouth is the anterior opening of gut. During development the entoderm in the anteroneurtral part of the phanjugeal region of the fore gut pushes out an evagination toward the ectoderm. The ectoderm also invaginates towards this evagination to form the stomodaeum already noted. The stomodaeal wall now meets and fuses with the endodermal wall in this region forming the *oral plate* or *oral membrane*. A few days after hatching (about 9 mm.), the oral plate becomes perforated, and hence forth the mouth communicates freely with the pharynix. Thus we see that the ectoderm plus the oral endoderm form the *oral plate*, or *oral membrane*, which generally ruptures through shortly after the time of hatching (6 mm. stage) to form the mouth opening. The lateral margins of the mouth (stomodeum) are the original *mandibular ridged* between which are the dorsal and the ventral (larval) lips. These are transitory but important feeding organs of the tadpole. The *dorsal lip* develops three medially placed rows of superficial *teeth* which are periodically sloughed off and replaced. The *ventral lip* of the larva develops four rows of somewhat more complete teeth, but all teeth are covered with stomodeal ectoderm. A cornified ectodermal "jaw," consisting of a hardened ridge, develops at the base of both the dorsal and the ventral lips.

By the time of metamorphosis the horny *larval teeth* and jaws are lost and the mandibular arches give rise to the jaw elements of the adult. The upper larval teeth are replaced by *permanent teeth* which bear only superficial resemblance to the teeth of mammals.

The tongue, which is ultimately attached anteriorly and will be free posteriorly, begins to develop by the time of metamorphosis from a proliferation of cells from the endodermal floor of the pharynx. The bulk of the tongue will, of course, be of mesodermal origin but most of its covering and glandular elements come from the endoderm, the anterior portion being covered with stomodeal ectoderm.

The Foregut

This anterior portion of the original archenteron expands widely in front of the yolk mass and consists of all of the derivatives from the stomodeum to and including the pancreas and liver. The stomodeum, the anterior covering of the tongue, and the roof of the mouth anterior to the internal choanae are ectodermal and only the posterior covering of the tongue and mouth, beginning at about the level of the thyroid gland, is endodermal. This foregut portion of the archenteron therefore assumes a major role in the early development of the frog, where it gives rise to three successive sets of respiratory organs (i.e., external and internal gills, then the lungs) to most of the endocrine organs, and to a good portion of the digestive tract.

The Pharynx

The *pharynegeal cavity* is widely expanded anterior and ventral to the yolk mass of the frog embryo. The earliest elongated and vertical evaginations of the pharyngeal endoderm are known as the *visceral* or branchial *pouches*.Those of the most anterior pair, which develop between the mandibular and the hyoid arches, are known as the *hyomandibular pouches*. The dorsal remnant of this first pair of pouches gives rise to the Eustachian tube of the adult and connects the pharynx with the middle ear chamber. The second and third pairs of visceral pouches are known as the first and second branchial pouches,

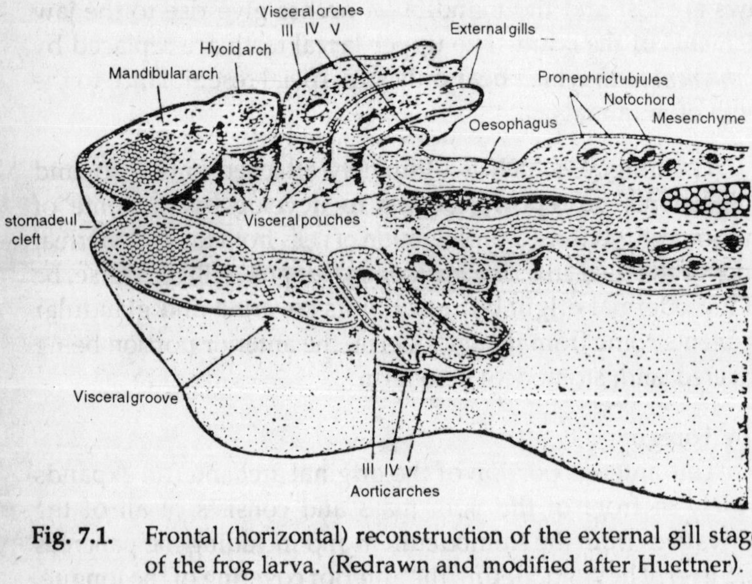

Fig. 7.1. Frontal (horizontal) reconstruction of the external gill stage of the frog larva. (Redrawn and modified after Huettner).

because they give rise to gill clefts, and they develop between the succeeding visceral arches. Eventually, six pairs of endodermal evaginations develop in this manner, but those of the last (most posterior) pair rarely open as pouches, remaining as mere cords of endoderm. By definition we use the term "pouch" for the endodermal evagination, "groove" for the ectodermal invagination, and "cleft" for the combination of ectoderm and endoderm which meet when these grooves and pouches break through. *Visceral pouches* I to V inclusive appear in this order and grow laterally toward the head ectoderm, all but the first and fifth, meeting a corresponding groove and breaking through to form clefts. Occasionally a sixth visceral pouch is developed, but it is generally vestigial.

These clefts can be classified in tabular form as follows:

Pouch (Endoderm)	*Arch (Mesoderm)* *Cleft (Ectoderm and Endoderm)*	*Groove (Ectoderm)*
	Visceral arch I - mandibular (Aortic arch I)	
	***Visceral cleft I** (Hyomandibular —	
Visceral pouch I	Eustachian tube Ectodermal plate— tympanic membrane	**Visceral groove I**
*Does not perforate		
	Visceral arch II — hyoid (Aortic arch II) (Oppercular fold arises at 9 mm stage, covers gills at 10 mm. stage)	
Visceral pouch II (Branchial pouch I)	**Visceral cleft II** (Branchial cleft I)	**Visceral groove II** (Branchial groove I)
	Visceral arch III (Branchial arch I) (Aortic arch III — carotid) (External gill at 5 mm stage)	
Visceral pouch III (Branchial pouch II)	**Visceral cleft III** (Branchial cleft II)	**Visceral groove III** (Branchial groove II)
	Visceral arch IV (Branchial arch II) (Aortic arch IV — systemic) (External gill II at 5 mm. stage)	

Visceral pouch IV (Branchial pouch III)	**Visceral cleft IV** (Branchial cleft III) *Visceral arch V* (Branchial arch III) (Aortic arch V) (External gill III at 6 mm stage)	**Visceral groove IV** (Branchial groove III)
Visceral pouch V (Branchial pouch IV)	**Visceral cleft V** (Branchial cleft IV) **Visceral arch VI** (Branchial arch IV) (Aortic arch VI — pulmonary)	**Visceral groove V** (Branchial groove IV)

The first four pairs of branchial clefts constitute channels or gill slits from the pharynx to the exterior, and are lined with both ectoderm and endoderm. The first visceral (hyomandibular) cleft and the sixth visceral cleft are rudimentary in that they rarely open through too the outside.

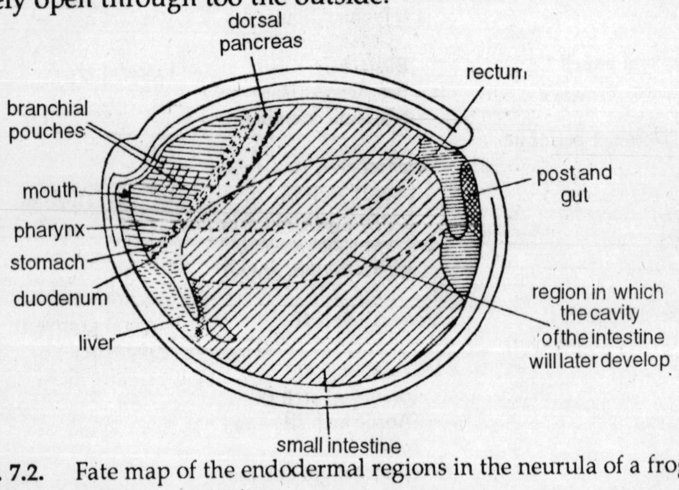

Fig. 7.2. Fate map of the endodermal regions in the neurula of a frog (Median section).

The finger-like and branched *external gills* arise as outgrowths of the lateral wall of the third, fourth, and fifth visceral (branchial I, II, III) arches shortly after the gill clefts are perforated. The first and second external gills arise at the 5 mm stage and the third at the 6 mm stage of development. The arches carry with them the covering ectoderm and the capillary

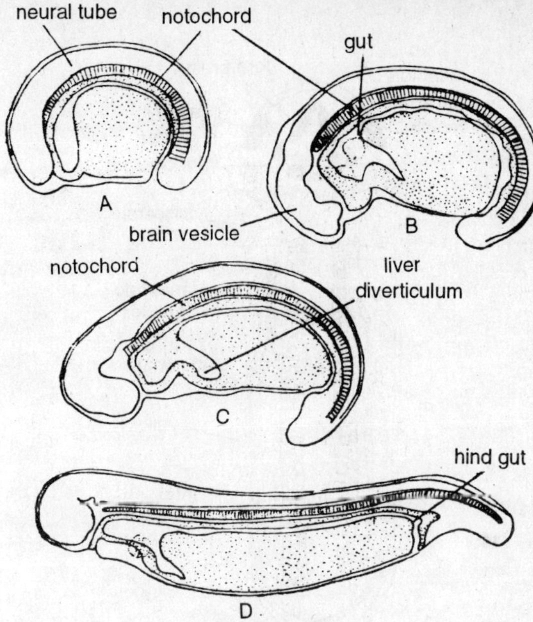

Fig. 7.3. Development of the endodermal organs in the amphibian (newt), from the neurula stage to swimming larva (median section).

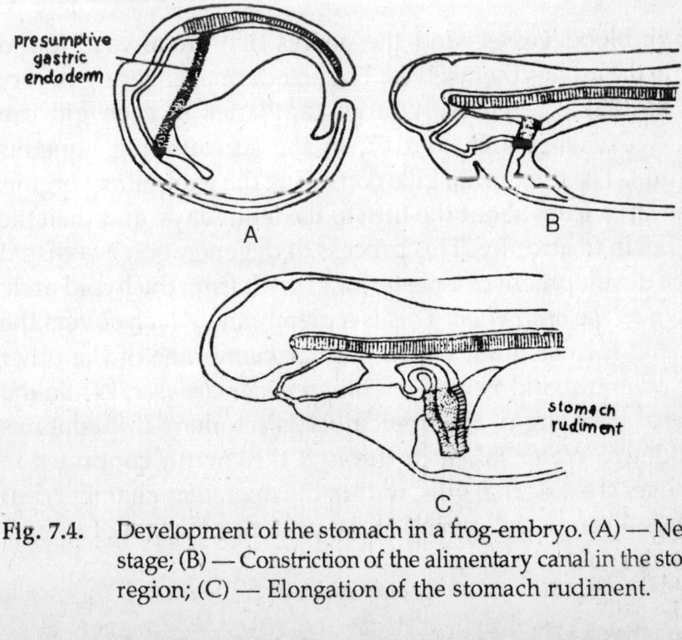

Fig. 7.4. Development of the stomach in a frog-embryo. (A) — Neurula stage; (B) — Constriction of the alimentary canal in the stomach region; (C) — Elongation of the stomach rudiment.

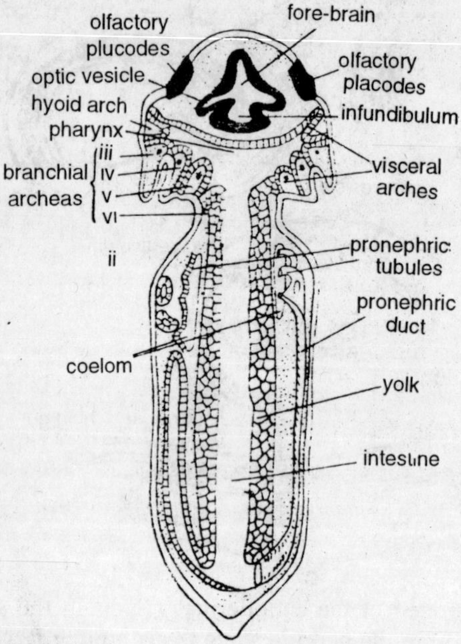

Fig. 7.5. Frontal section of tadpole shortly after hatching.

loop of blood vessels and the nerves that are always found within the arches themselves. The ectodermal covering is very thin so that the relatively large capillaries in each gill can readily exchange CO_2 and O_2 in the surrounding aqueous medium. These external gills constitute the respiratory organs of the larva from about the fifth to the tenth days, and then the gills begin to atrophy. This process of degeneration is assisted by the development of a posterior growth from the hyoid arch, known as the *operculum*. This is a membrane which covers the gills and forms, along with a similar membrane of the other side, a ventral and ventro-lateral *opercular chamber*. While the external covering of the operculum is ectodermal it contains mesoderm. Water taken in through the mouth continues to pass over the external gills, within the opercular chamber, but escapes through the *spiracle* at the posterior margin of the left opercular fold.

In the meantime the *internal gills* must develop in order to take over the respiratory functions that gradually are being relinquished by the external gills. These internal gills develop a double row of filaments, ventral to the branchial arches and arising doubly from the postero-external (anterior and posterior) faces of the same first three pairs of branchial arches. These are visceral arches III, IV and V. The fourth pair of branchial (sixth visceral) arches may also give rise to reduced single internal gills from their anterior faces only. These gills are termed internal because of their position on the arches and also because of the fact that they are covered by a body flap, the operculum, and are therefore truly within the body. The entire opercular (branchial) chamber is lined with ectoderm, however, and this includes the covering of all of the external and internal gills. The internal position of this new set of gill filaments is therefore secondary.

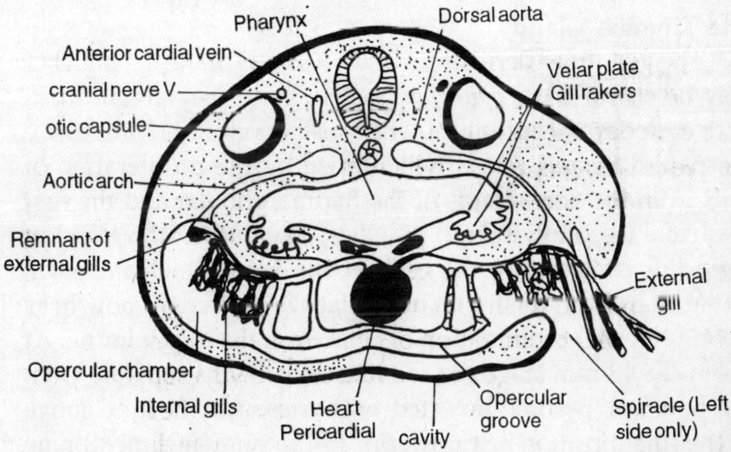

Fig. 7.6. Relation of the pharynx to the internal and external gills of the frog, transverse section.

The ventro-lateral position of the external gills, plus the development of the operculum, tends to move them to a position beneath rather than lateral to the pharynx, within the spacious *opercular cavity or chamber*. The original endodermal branchial pouches, lateral to the pharynx, become partially separated off by lateral projections of the pharyngeal floor and also by longitudinal folds in the lateral aspects of the roof. This provides a lengthwise pair of pockets between the pharynx and the latero-ventral gill chambers. From the floor of each of these pockets develop finger-like, endodermally covered, frilled organs known as the gill rakers. These tend to filter the water as it passes from pharynx to gill chamber, removing any relatively large particles of matter and retaining them within the pharynx to be carried into the oesophagus with the food. In addition, the shelf-like projection or flap of tissue from the floor of the pharynx and also from the lateral wall of the pharynx likewise aid in the mechanical sifting or filtering of the water. These shelves are known as the *velar plates.* During subsequent metamorphosis both the gill chamber and the related opercular chamber become filled with rapidly multiplying cells which ultimately become incorporated into the body wall.

The Thymus Gland

The glandular derivatives of the various branchial pouches may be classified as *epithelioid bodies,* since they are all lined with endodermal epithlium. The most anterior derivatives are the paired *thymus glands* which arise by the proliferation of cells from the dorsal ends of the hyomandibular and the first branchial (second visceral) pouches. The bulk of the gland of the adult comes from the cells of the first branchial pouch, rather than from the hyomandibular (first visceral) pouch, as a solid internal proliferation of cells from the upper lining. At about the 12 mm stage the sac-like outgrowths separate from the pouches, become invested with mesenchyme, and move to the final position just posterior to the tympanic membrane and ventral to the depressor mandibulae muscle. This gland is larger in the younger frogs, attaining its maximum size in frogs of about 20 mm body length. It is a lymphoid gland, presumably a source of some blood corpuscles, and apparently is essential to the early development of the frog.

The Carotid Glands

From the ventral ends of the first branchial (second visceral) pouch there arise cell proliferations, at about the time the internal gills appear (9 mm stage) which develop into the *carotid glands*. These glands develop and move to the junction of the internal and external carotid arteries. Their function is to regulate the flow of blood, particularly that entering the internal carotid artery. This is accomplished by means of its final spongy consistency. They may aid also in achieving adequate aeration of the blood.

The Parathyroid Glands (Pseudo-thyroid or "Ventraler Krimenust")

From the ventral ends of the second and the third branchial (third and fourth visceral) pouches are derived the *parathyroid bodies*, otherwise known as the pseudo-thyroid bodies. In the adult these are small, rounded, vascular glands which come to lie on either side of the posterior portion of the hyoid cartilage.

The Ultimobranchial Bodies (Supra-pericardial or Post-branchial Bodies)

The fourth pair of branchial (fifth visceral) pouches have no glandular derivatives, but a pair of *ultimobranchial* (post-branchial) bodies arise by cellular proliferations from the rudimentary fifth branchial (sixth visceral) pouches. These organs arise as solid proliferations of the pharyngeal wall and come to lie beneath the mucous membrane of the pharynx lateral to the glottis. They shortly separate from the pharynx and acquire cavities. The structure in the adult is somewhat like that of the thyroid gland but the function has not yet been determined.

The Thyroid Gland

The thyroid is an endocrine gland which arises as a single median thickening and evagination from the floor of the pharynx between the base of the second pair of visceral arches, just before the time of hatching and at about the 5 mm stage. It late becomes a bi-lobed and solid organ, far removed from its site of origin. Only its original lining endodermal, the bulk of the

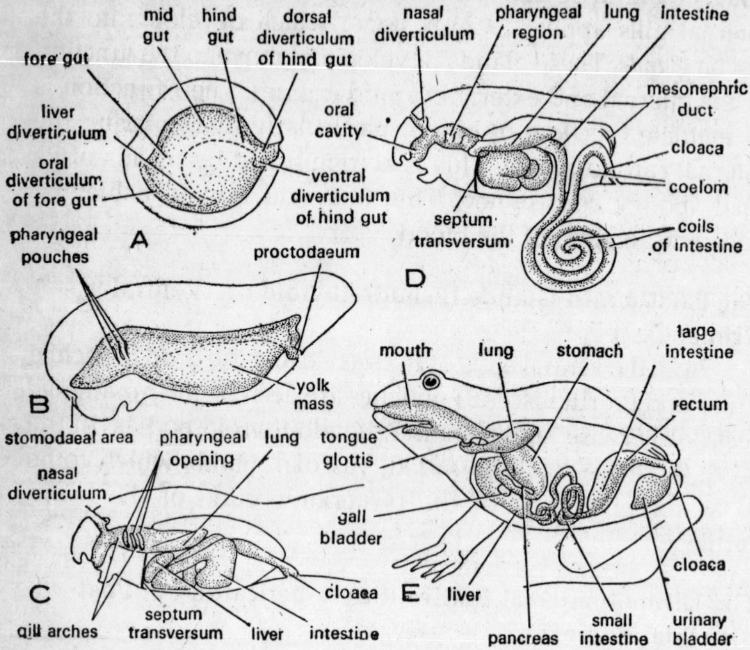

Fig. 7.7. Morphogenesis of the digestive tract in the frog. A — Early postgastrula. B — 4 mm tadpole. C — 9-10 mm tadpole. D — 15 mm tadpole. E — Adult frog.

gland being mesodermal in origin. At about the 10 mm stage, however, it becomes separated from the pharyngeal floor as a close sac. This divides into two lobular and vesicular masses, and moves to a position on either side of the hyoid cartilage apparatus. The thyroid gland of the 15 mm stage is divided but retains a connection at its anterior ends by a short isthmus, so that it takes the shape of an inverted "Y," with a short base. The wings move posteriorly, along the ventral face of the hyoid cartilage, and these changes are correlated with gradual changes in the hyoid apparatus. The glands enlarge, the surrounding cartilaginous elements expand, the major blood vessels become enmeshed in them, and eventually they come to lie close to the heart. At the forelimb emergence stage of metamorphosis there is a marked distension of the follicles, and by the tail resorption stage there is high epithelium liquefaction while erosion of the colloid and follicular collapse occur. At this time the genio-hyoid, hypoglossus, and sterno-hyoid muscles are in close proximity to the thyroid, a situation which does not persist to the adult. The gland shows hyperactivity during metamorphosis with relative inactivity at the completion of the metamorphic process. Its activity during these phases of development is closely correlated with the development and function of the pituitary gland, particularly of the basophilic cells in the pituitary gland.

In later stages it may be seen attached to the ventral aspect of the hypoglossus muscle. The fully formed thyroid gland consists of separate follicles, each made up of a single circular layer of cuboidal (endodermal) epithelial cells, in the center of which is a lumen filled with a colloidal mass.

The Tongue

The tongue appears just before metamorphosis and is indicated as an elevation in the anterior floor of the pharynx, just posterior to the site of origin of the thyroid gland. Anterior to this the pharyngeal floor is depressed and glandular, but during metamorphosis this glandular area becomes the free anterior tip of the tongue.

The Lungs

Shortly after the time of hatching, when the larva measures about 6 mm in length, there appear bifurcating but solid cell proliferations from the pharyngeal endoderm just behind the developing heart. These soon develop into paired saccular evaginations, directed posteriorly. These *lung buds* arise from the median ventral floor of the pharynx at about the level of the rudimentary sixth visceral (fifth branchial) pouch. The single short tubular connection of the lung buds opens into the foregut through the *glottis*. The connecting column of cells which has acquired a lumen, from which the lung buds arise, will become the *trachea*. At the level of the glottis there is a short transverse chamber known as the *laryngeal chamber*. The more posterior bi-lobed mass eventually will open up as the *primary bronchi* or lung buds. In the 11 mm stage each lung bud will be surrounded by peritoneal epithelium and will be invested with splanchnic mesenchyme. Later, each lung will become on ovoid, thin-walled, and slightly alveolated sac, lined with endodermal squamous epithelium. Outside of this epithelium, and constituting the substance of the lung, are connective tissue, blood, and lymph vessels, all of mesodermal origin.

The Liver

The liver originates very early as a single median ventral endodermal diverticulum which is directed posteriorly between the heart rudiment and the yolk mass. The diverticulum enlarges slowly and its anterior wall will thicken, become folded, and finally branched, to form the liver proper. The ultimate lobes of the liver will retain their tubular connection with the original diverticulum as the *hepatic duct*. The original diverticulum will elongate as the *bile duct* leading to a terminal vesicle, the gallbladder, which becomes very large in the tadpole. All of these derivatives become invested with connective tissue and blood from the splanchnic mesoderm, but some of the adjoining yolk cells become the true hepatic cells.

The Pancreas

The pancreas arises as three rudiments at about the 9 mm. stage in a manner somewhat similar to the liver. The first rudiment appears as a single posteriorly directed ventral diverticulum of the bile duct at its point of entrance into the foregut. This diverticulum soon divides into two and the cellular elements grow around the bile duct to fuse into a single mass of spongy tissue anterior to the bile duct. Subsequently a third mass of similar tissue arises from the dorsal wall of the gut, and attains junction with the original masses, all three to form the much-lobulated pancreas. These three pancreatic rudiments then use the single *pancreatic duct* which retains its original connection with the gut, just posterior to the liver or hepatic duct. It marks the boundary between the foregut and the mudgut. The cells of the islet of Langerhan arise early from endoderm.

The Oesophagus and Stomach

After hatching, the undifferentiated portion of the foregut between the glottis and the bile duct elongates to become the *oesophagus*. During early larval development this part of the gut becomes temporarily occluded by an *oesophageal plug* of cells whose origin is unknown. Its function during its brief appearance may be to help direct any water and food from the mouth out over the gills. These early larvae require no external source of food as they are supplied with abundant yolk. The oesophageal plug disappears by the 9 to 10 mm stage of development.

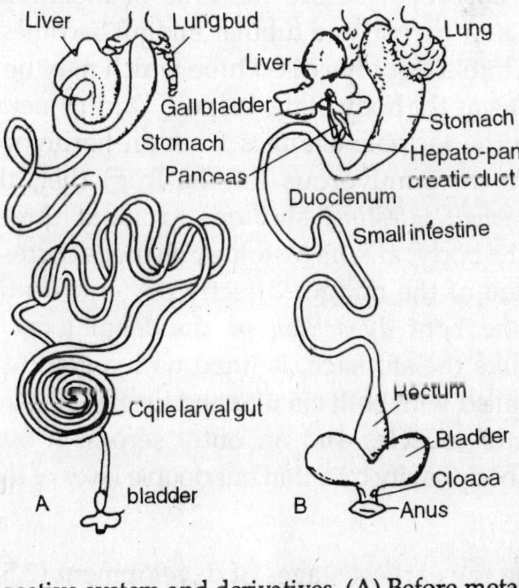

Fig. 7.8. Digestive system and derivatives. (A) Before metamorphosis. (B) After metamorphosis. (Modified and redrawn from the Leuckart wall chart).

Beginning at about the 11 mm stage, the oesophagus develops into the very distensible organ of the adult while the *stomach* becomes differentiated simply by a dilation of the next adjacent part of the foregut. By the time of metamorphosis the stomach is somewhat more distended than the oesophagus but it is otherwise indistinguishable from it. In the newly metamorphosed frog, however, it assumes a transverse position, shifting from the earlier longitudinal position, and has all the characteristic layers of any vertebrate gut. This includes an inner and glandular layer of mucosa, intermediate layers of connective and muscular tissue, and an outer covering of serosa.

The Midgut

The midgut is that portion of the original archenteron which is found dorsal to the yolk mass as long as this mass exists, having a roof and sides one cell in thickness. The floor is the thick yolk endoderm. After the time of hatching the yolk is rapidly absorbed. Before the time of metamorphosis the heretofore undifferentiated tubular midgut becomes very much elongated into a doubly-coiled tube which may be nine to ten times as long as the body of the tadpole. During metamorphosis, and while the tadpole changes from an herbivorous (of the tadpole) to an omnivorous (of the frog) diet, this midgut (potential *small intestine)* shortens to about three times the length of the body, and its histology changes correspondingly. That portion of the midgut directly posterior to the stomach becomes the bent *duodenum* or duodenal loop. The small intestine, like the stomach, is lined with a glandular mucosa and is supplied with both circular and longitudinal (involuntary and smooth) muscles and an outer serosa. It is suspended within the body cavity by a thin but double layer of the peritoneal epithlium.

During the earliest stages of development (2.5 mm stage) there may be seen a *sub-notochordal* or *hypochordal rod* of pigmented cells, two or three cells in diameter, lying between the roof of the midgut and the notochord. There is positional and structural evidence that this column of cells is at some time associated with the roof of the archenteron, from the level of the liver to the posteriorly placed neurenteric canal. It becomes entirely free from gut and notochord by the 4.5 mm stage and disappears shortly after the time of hatching. It has no known function. It may be of evolutionary and ontogenetic significance only.

The Hindgut

This is the smallest portion of the original archenteron which lies posterior to the yolk mass, between it and the posterior body wall. The endoderm ventral to the closed blastopore evaginates to fuse with the invaginating proctodeal ectoderm to form the *anal plate*. This plate finally ruptures at about the 4 mm stage to form the anus, only the inner portion of which is lined with endoderm. The *rectum* is the enlarged posterior end of the archenteron, and is therefore endodermally lined. This does not develop until the time of metamorphosis.

Dorsal to the rectum are a temporary extension of the archenteron, developed in consequence of the presence of the *neurenteric canal*, and the posterior growth of the notochord. This is the post-anal gut. At the 5 mm stage only a remnant of the *neurenteric canal* can be identified as parallel rows of pigmented cells extending ventro-posteriorly from the hindgut. This is the region of the disappearing postanal gut. It represents the enteric portion of the temporary neurenteric canal connecting the posterior ends of the neurocoel and archenteron. It has no derivative in the adult, but has homologues in the development of most of the vertebrates.

That portion of the hindgut between the rectum and the anus forms an ectodermally lined chamber known as the cloaca. Into this chamber empty the paired urogenital ducts, to be developed later. Before metamorphosis there appears a ventral evagination of the cloaca which gives rise to the bladder. This ultimately bi-lobed and elastic organ has no direct connection with the excretory ducts as do the bladders of higher forms. However, it is considered to be a reservoir, by way of the cloaca, for excretory fluids. In the closely related toads, which live in hot sand, it also may be a reservoir for water storage and an aid in respiration.

8

Embryonic Determination or Induction

As in the chapter on sea urchin development, an attempt will be made here to summarize the various approaches to the problem of differentiation in the amphibian egg. Experimentation has proceeded from the level of tissues and cells to the molecular level. Development of new experimental techniques and approaches depends on technical advances and also on fundamental discoveries in the field of biochemistry. In recent years, the most intensive effort has been at the molecular level. The problems are the same as those encountered in the egg of the sea urchin: the nature and origin of egg polarity; the capabilities of various parts for self-differentiation; the interdependence of different regions upon each other. At the molecular level, the analysis has been mainly concerned with a description of the normal sequence of biochemical events during development of the embryo, most particularly with the synthesis of DNA and RNA and protein during development. The study of transcription and translation of genetic information again provides knowledge of the relationship between gene-expression and differentiation.

Isolation of Blastomeres; Evidence of a Regulative System

The first experiment on embryonic material of any kind is generally conceded to be that of Wilhelm Roux who studied the effect of killing one of the first two blastomeres of the frog egg. The results of this study were published in 1888. The experiment is of considerable historic interest, even though the conclusions reached by Roux turned out to be in error. Roux was the first biologist to work in the area of causal analysis of development. His underlying motive was to resolve problems of development into fundamental processes which he assumed to be the same fundamental processes underlying inorganic phenomena. This was a strong departure from the ideas of the day, which were mainly concerned with explanations of embryological phenomena in terms of comparative anatomy or evolutionary data. Prior to Roux, an embryonic structure could be considered to appear because it was part of the phylogenetic history of the species to which the embryo belonged. Roux formally put forth the idea that the proper study of the cause and effect during embryonic development was the embryo itself. He founded a new discipline of causal, analytical embryology which he called developmental mechanics. He is frequently called the "father of experimental embryology."

The experiment reported in 1888 was simple in design and was intended to decide the issue of independent versus interdependent (mosaic versus regulative, respectively) development as the underlying principle of differentiation in the frog egg. The experiment was performed at the two-cell stage and consisted of injuring one of the first two blastomeres after the first cleavage. Since the surviving cell gave rise to half an embryo, Roux felt that each cell developed independently of its neighbour and that development of the whole embryo represented the summation of the development of each part. These conclusions were later shown to be erroneous for the frog egg. If the first two blastomeres were isolated from each other (instead of leaving an injured cell in contact with a live one), each blastomere was found to develop into a complete tadpole and frog, on most occasions. Even though Roux's experiment was shown to be of faulty design, the foundations were laid for the new experimental approach to the study of embryology.

Later experiments also showed that two complete embryos

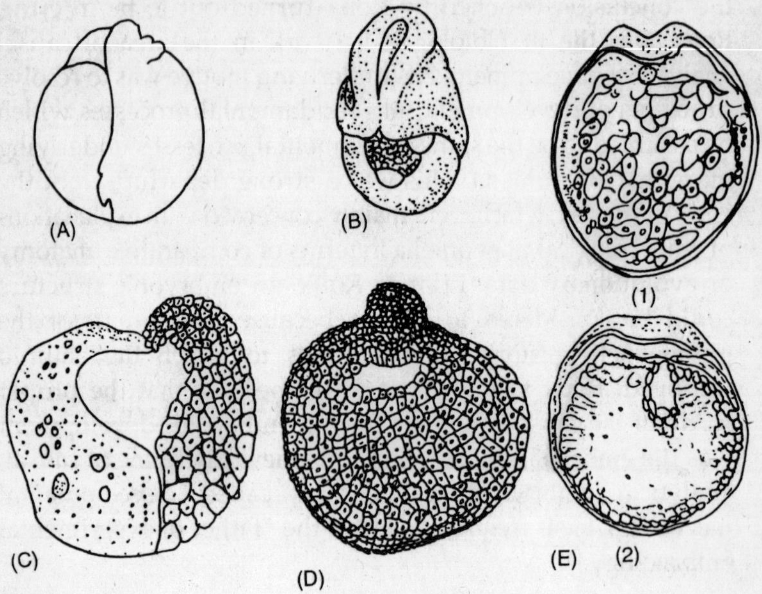

Fig. 8.1. (A) Half embryo of frog, resulting from injury to one of the first two blastomeres; (B) anterior embryo, so-called, after injury to one of the first two blastomeres of an egg in which the first cleavage had been at right angles to median plane. This embryo may also be interpreted as a whole embryo in which the lateral lips of the blastopore have been prevented from coming together by the material of the injured half; (C) section of blastula stage of $^{1}/_{2}$ blastomere; (D) section of $^{1}/_{2}$ embryo like that shown in A. (E) Cross sections through dorsal embryos arising from isolated *Triturus* blastomeres. (1) Half embryo; (2) quarter embryo, some magnification, (dorsal meaning the gray crescent material was included).

resulted from isolation of the first two blastomeres only when the first cleavage occurred in a plane that bisected the gray crescent area. In those instances where the first cleavage plane occurred at some angle to the plane of bilateral symmetry instead of parallel to it, only the blastomere containing the gray crescent material developed into a complete embryo as diagrammed in Figure 8.1 The blastomere that lacked the gray crescent developed only into an undifferentiated mass of epidermal cells. These results, along with other studies, served to focus attention on the gray crescent area in later experimental work.

Fusion of Two Embryos; Additional Evidence of Regulative Ability

Two whole amphibian eggs can be merged to form a single embryo of double size. Two 2-cell stages are cross united in the manner. The result is regulation to form a single embryo, not a double "monster." Again, this result confirms the theory of interdependence of developing parts rather than theory of independent, self-differentiation of parts.

Experiments on the Gastrula; Interchangeability of Presumptive Neural Plate and Epidermis

The early gastrula, as shown by the fate map, has a roof over the blastocoel composed of cells destined to give rise to neural plate and epidermis. Prior to extensive involution of the presumptive chordamesoderm at the dorsal lip of the blastopore, the roof of the blastocoel can be rotated 180° such that presumptive neural plate and epidermis have exchanged locations. The result of this operation is a completely normal embryo indicating that, at this time, the two tissues are interchangeable (regulative).

Inductive interaction between cells is shown by transplants of the dorsal lip

The developmental fate of the cells turning in at the dorsal lip of the blastopore is largely chordamesoderm. If the chordamesoderm is transplanted to the ventral epidermis of an early gastrula host embryo, the transplant tissue will turn under and come to lie beneath the ventral epidermis of the host embryo. At this point an entire secondary neural axis will arise on the ventral side of the host. This experiment was first performed by Spemann and Mangold, who went on to demonstrate that the neural tube of the secondary embryo is of host origin and the mesoderm is of graft origin. The Spemann

— Mangold experiment involved the use of two salamander species of differing pigmentation as donor and host tissues. Sipenann gave the dorsal lip of the blastopore the functional designation of inmary organizer," reflecting its role in the organization of the neural axis. The presumptive chordamesoderm, upon transplantation or explanation to culture medium, is capable of self-differentiation into those structures assigned to the dorsal lip by fate map studies.

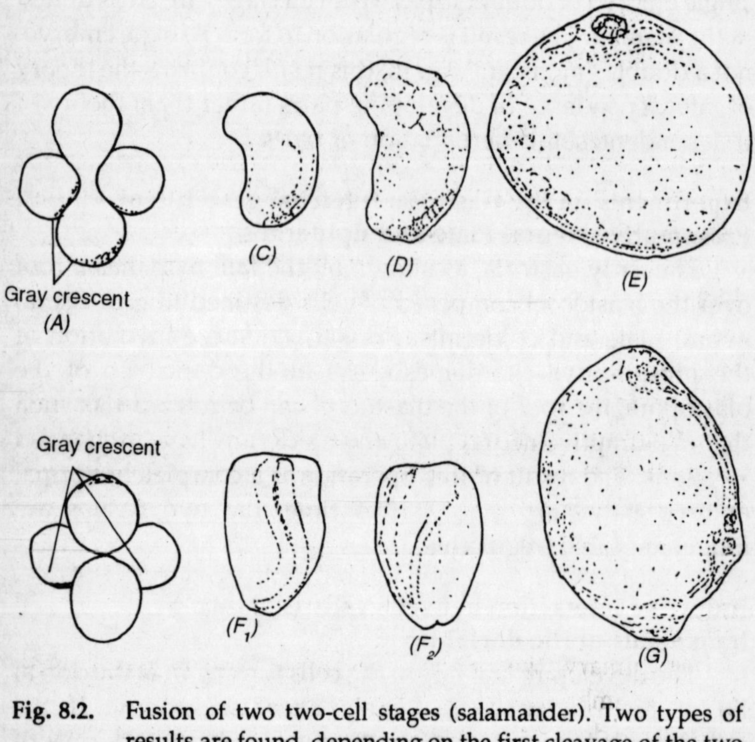

Fig. 8.2. Fusion of two two-cell stages (salamander). Two types of results are found, depending on the first cleavage of the two fused eggs.

The action of the dorsal lip is apparently, from the experiment above, to organize the overlying presumptive neural plate into the neural axis. This is borne out by the following observations:

1. The neural axis fails to develop in those embryos whose primary organizer regions have been removed.

2. Accessory neural axes are induced in overlying presumptive epidermis in the presence of implanted dorsal blastoporal lip material.
3. The presumptive epidermal and neural areas are interchangeable in the early gastrula, as shown by experiments involving rotation of the roof of the blastocoel through 180°.
4. In embryos induced to form exogastrulae, in which the chordamesoderm self-differentiates but does not come to lie beneath the presumptive neural plate cells, fail to differentiate as neural cells (Figure 8.4).

The Chordamesoderm is Regionally Specific

Transplant studies show that posterior regions of the chordamesoderm induce posterior neural structures; anterior portions to form. This can be shown by defect experiments, in which defects are made in the chordamesoderm at different times during gastrulation. Defects are thus caused at different levels in the chordamesoderm and produce corresponding defects in the overlying neural tube. The action of the dorsal lip material (primary organizer) is an example of embryonic induction, in which an inductor tissue evokes a response (differentiation) in tissue competent to respond to the stimulus. This competence extends to presumptive neura plate areas and to presumptive epidermis. Competence to respond to the primary organizer is lost in later (neurula) stages.

In summary, the primary organizer induces the neural axis of the embryo. The organizer activity is confined to the region of the dorsal lip of the blastopore. The organizer is self-differentiating and regionally specific in its action. The location of the dorsal lip and, therefore, the organizer corresponds to the location of the gray crescent material of the uncleaved, fertilized egg.

Cortical Transplants Provide Evidence for Regionalization of Egg Cytoplasms

As described earlier, the gray crescent is a cortical area

characterized by a retreat of pigment from the equator at fertilization. It lies at the boundary between the black and white hemispheres of the frog egg and marks the future dorsoposterior side of the embryo. It will be re-called that the gray crescent region of the fertilized frog egg corresponds to the region of the primary organizer and the chordamesoderm. Development of isolated blastomeres is complete is both isolated cells only if both contain gray crescent material in nearly equal amounts. These observations focus attention on the gray crescent cortex of the egg as the initial organizer center.

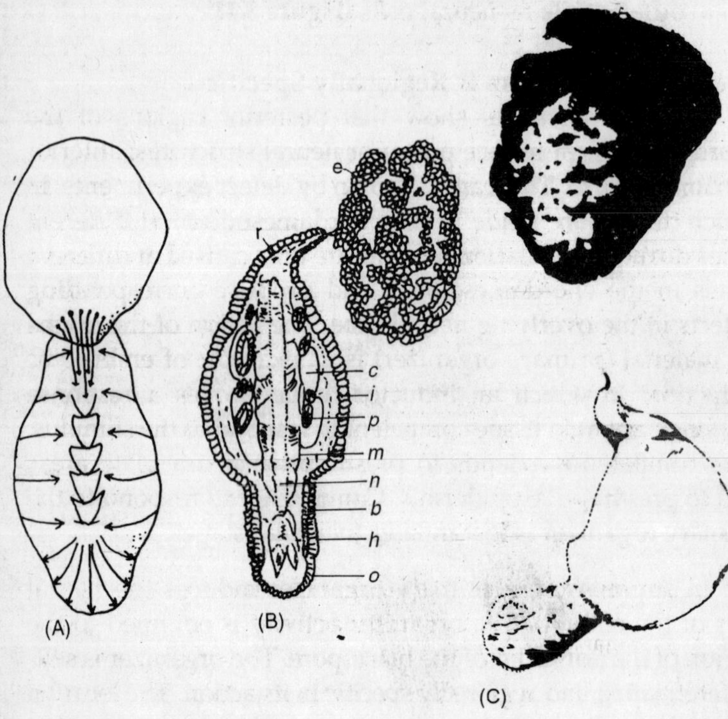

Fig. 8.3. Complete exogastrulation of a salamander germ.

Recently it has appeared possible to transplant or remove the cortical gray crescent material. The results of these operations are summarized. The gray crescent material behaves in much

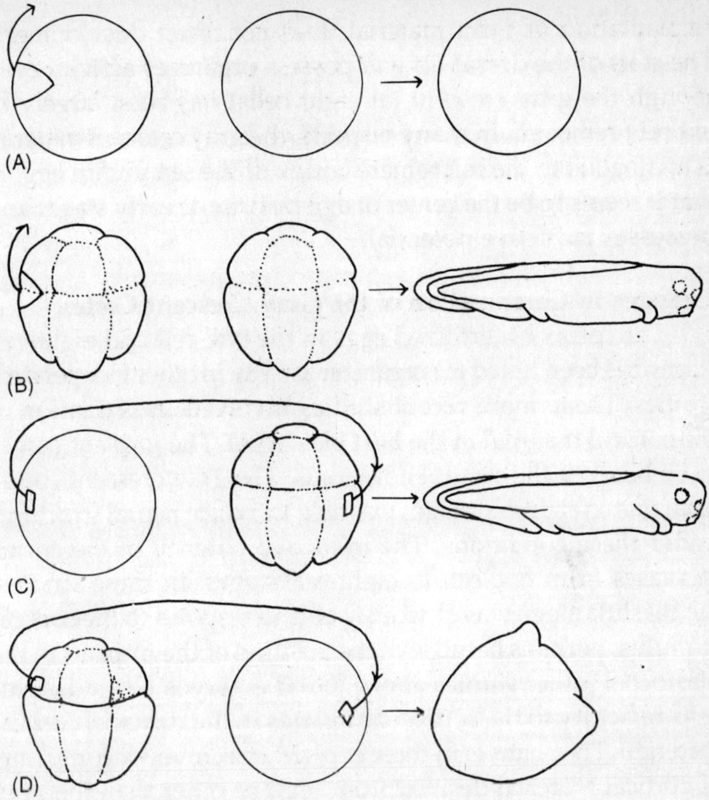

Fig. 8.4. Cortical transplant in *Xenopus.* (A) Excision of the cortical gray crescent area the one-cell stage; no gastrulation; (B) same experiment at the eight-cell stage; normal embryo; (C) graft of the gray crescent cortex of the one-cell stage to the ventral part of the eight-cell stage does not result in the induction of a secondary embryonic axis; (D) graft of the gray crescent cortex from the eight-cell stage to the ventral margin of one cell stage a secondary embryonic axis.

the same way as the dorsal lip material with regard to neural induction. Removal of the gray crescent (up to the eight-cell stage) terminates neural development. Prior to the eight-cell stage, implanting an extra gray crescent produces an extra neural axis. The gray crescent cortex has apparently discharged its function by the eight-cell stage, after which removal or implantation of extra material does not affect development. The cells of the dorsal lip will possess organizer activity even though the gray crescent (at eight cells) has been largely or entirely removed. In many respects, the gray crescent material is analogous to the micromere cortex of the sea urchin egg, in that it seems to be the center of egg polarity at early stages and possesses inductive potential.

Changes in Composition of the Gray Crescent Cortex

The cortex of fertilized eggs in the one-cell and eight-cell stages has been tested for organizer activity in other independent studies. These more recent studies involved implantation of the cortical material in the host blastocoel. The implant comes to lie beneath the ventral epidermis. The gray crescent cortex from one-to eight-cell stages was able to induce neural structures under these conditions. The inductive potency of the cortex increases from one-cell to eight-cell stages. In these studies, the blastula stage was obviously able to respond to the cortical stimulus, perhaps because of the position of the implant in the blastocoel (other studies above found embryos beyond eight-cells refactory to the action of implants in the cortex of cortical material). The controls in these experiments, involving implants of cortical material derived from regions other than the gray crescent cortex, were uniformly negative. This indicates the activity is localized in the gray crescent cortex. Further examination, using micro-electrophoresis techniques, allowed study of the protein components of the cortical areas. The cortex was found to be regionally different in composition (the gray crescent differs from other regions) and it was found that the composition within the gray crescent cortical are changes between the one-celled and eight-celled stages.

It is tempting to assign a causal role to the proteins found in association with the gray crescent, since differences in localized protein composition are correlated with differences in morphogenetic activity. Unfortunately, there is no evidence yet to support or negate such speculation.

It has been confirmed, from two sources, that the gray crescent cortex contains information for axiation and neuralization of the embryonic ectoderm.

Chemical Nature of the Primary Organizer

That living cells are not necessary for primary induction is shown by the ability of killed (boiled) dorsal lip material to induce neural structures when transplanted beneath the ventral (competent) epidermis of an early gastrula host. Induction may also take place across a non-living membrane which allows the passage of macromolecules but not cells. However,

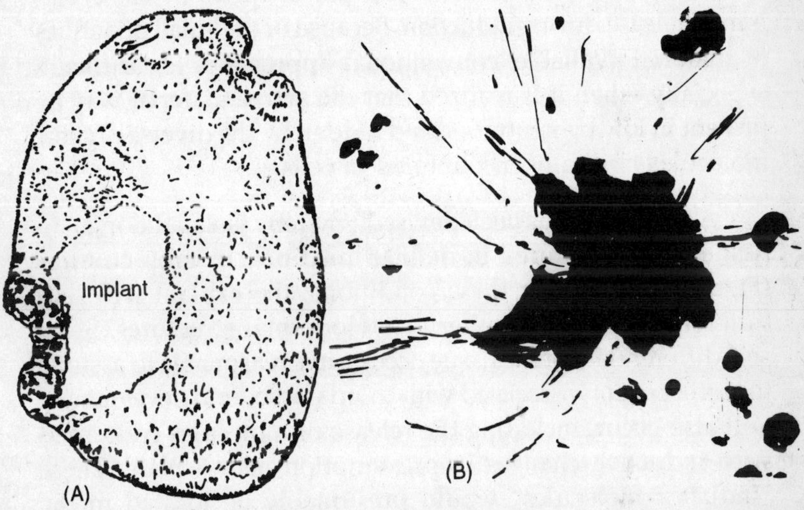

Fig. 8.5. (A) Induction of secondary neural plate in an axolotl embryo by means of an implanted piece of agar containing substances extracted from muscle by boiling the muscle in water. (B) Neural tissue with radiating nerve fibers and melanophores developed from a piece of young *Triturus ricularis* ectoderm cultivated in conditional medium.

it has been difficult to establish that induction across membranes of known porosity does not involve the intercellular contact of fine cell processes which are able to insert themselves through the membrane pores. Induction will not take place across cellophane, which allows neither macromolecules nor cell extensions to penetrate. It is therefore uncertain whether or not normal induction involves the release and uptake of materials from inducer tissue to competent tissue, and whether or not cell contacts are required for induction under normal circumstances.

With the above restrictions in mind, the task of identifying the chemical naturally responsible for neural induction is a formidable one. Many extracts and substances have been tested and many were active. The problem is still to find a means of examining the natural inductor (the dorsal lip) for the molecules it may release during induction. Because of the small quantities of material available, conventional approaches are difficult, especially when it is realized that the active material may be present in low concentration and hidden by the diverse organic molecules normally encountered in cells.

A technique has been devised whereby isolated dorsal lip tissue can be cultured in defined medium, in drop cultures (Figure 8.6*B*). Under these conditions, the explant will self-differentiate into a number of mesodermal structures. Since self-differentiate of the tissue occurs, it is reasonable to assume that other events associated with chordamesoderm development will also occur, including the release of molecules concerned with inductive changes in presumption neural plate. These "inductor molecules" would presumably be located in the culture medium, since the dorsal lip cells were cultured without responsive cells. Using this system and attendant assumptions about the release of active molecules, the following biological assay was developed. Medium in which presumptive chordamesoderm had been growth for a number of days was recovered, free of cells. This medium contained the original inorganic components, plus any molecules, contributed by the differentiating chordamesoderm cells. In new cultures the chordamesoderm medium (termed conditioned medium) was used to culture pieces of presumptive epidermis. After some days of culture, the tissue was examined for evidence of neural development (presence of cells with nerve fiber outgrowths; chromatophores derived from neural crest cells. The presence of neuralization in conditioned medium and the absence of neutralization in control cultures prepared using medium conditioned not by chordamesoderm but by presumptive epidermal cells, were taken as evidence that the inductor molecules were present in the test-conditioned medium.

In practice, conditioned medium consistently produced cultures showing evidence of neutralization, while control cultures were negative. The conditioned medium was found to contain many biological molecules, among them nucleic acids and proteins, as revealed by the absorption spectrum of the conditional medium in ultraviolet light. The system can be used for finer analysis of the chemical nature of the inductor molecules. Specific enzymes can be employed to specifically inactivate different kinds of biological molecules. Proteolytic enzymes (trypsin) would be expected to remove inductor activity associated with proteins. Nucleases (RNase and DNase) would be expected to remove activity associated with nucleic acids. Actual experiment shows that only trypsin completely abolishes the inductor activity of conditioned medium, though RNase diminishes the activity without abolishing it completely. These experiments suggest that the material involved in the induction of the neural axis by the underlying chordamesoderm is protein in nature, possibly a ribonucleoprotein.

Inductions Using Foreign Materials

Obviously, the amount of tissue available for analysis in the dorsal lip of the blastopore is extremely limited. In the course of experiments on killed inductors (cell extracts, or heated preparations), it was found that mammalian tissues yielded preparations having inductive activity when tested on amphibian embryos. Thus, boiled guinea pig bone marrow caused the competent, early gastrula, presumptive epidermis to develop into spinocaudal structures (tail musculature and other mesodermal derivatives). Other tissues, such as guinea pig liver, yielded material causing the induction of anterior neural structures. The active materials all seem to be protein in nature. Later, it was discovered that active material could also be extracted from whole amphibian gastrulae.

These abnormal inductors have been highly purified and preliminary word has been done on the mode of action of the proteins. The spinocaudal factor (mesodermal or vegetalizing factor was inhibited by nucleic acids. The inhibition was due, at least in part to the binding of the factor to the nucleic acid (DNA or RNA). The source of the nucleic acid was not of consequence, chick DNA or RNA serving as well as amphibian DNA or RNA. This has been interpreted to mean that the inductors might be acting by processing RNA some way, facilitating its transport to cytoplasmic regions. These speculations have little to espouse them. There is little evidence for the activity of these factors in the normal inductive process. The situation is reminiscent of the situation encountered in artificial parthenogenesis. Many different agents can act to activate the egg. None is necessarily involved in normal fertilization processes. The ability of an "inductor" to alter the fate of a cell is of interest as a system of cell differentiation, but does little to elucidate what actually happens during the normal determination of the presumptive neural plate of the amphibian gastrulae.

Interestingly, these kinds of results have led some to theorize that induction involves changes in intracellular concentration of various ions or changes in *intracellular* pH. Lithium, sodium, or calcium can all induce differentiation of neurons from presumptive epidermis. Later treatment with the same ions also induces differentiation of these same neurons into pigment cells. The observations are supported by the occurrence of great changes in concentrations and proportions of cations *in vivo* during the induction of the nervous system.

The effects of the conditioned medium and other preparations are apparently not medicated through damage to the competent ectoderm cells, but through the interaction of the molecules with the cells. There is reason to believe that inductor molecules act indirectly on competent tissue to cause a response. The extent to which the abnormal inductors are identical to the natural substances from the dorsal lip is not known.

Inductive Action is Indirect and Mediated Through the Responsive Genome

These analyses involve the use of two species; one, the source of the inductor; and the other, the competent responding tissue. The response to the inductor is always in accord with the genetic information contained in the responding tissue. For example, salamander larvae possess a structure on either side of the head just behind the jaw angle and beneath the eye. This structure is called a balancer. Anuran embryos possess homologous structures beneath the mouth variously called mucous glands, adhesive organs, or oral suckers. These structures (balancers and mucous glands) are induced in overlying epidermis by inductive action of the anterior roof of the archenteron. Speaking first of balancers, there exist some species of salamander that lack these organs. If epidermis from any part of the body of a species normally possessing a balancer is transplanted to replace the balancer epidermis of a species not possessing a balancer, a balancer will develop. In the reverse experiment, epidermis from a species not possessing balancer fails to develop a balancer when placed over the balancer inductor in the species normally possessing balancers. The balancer inductor is obviously operative in both species. The ectoderm of the species lacking balancers is genetically not competent to respond to either inductor, whereas the epidermis from the species having balancer is competent in either location. The main point is that the genetic lack of competence is not overridden by the inductive stimulus which is present in both species. Inductors cannot alter the genetic nature of the responding tissues.

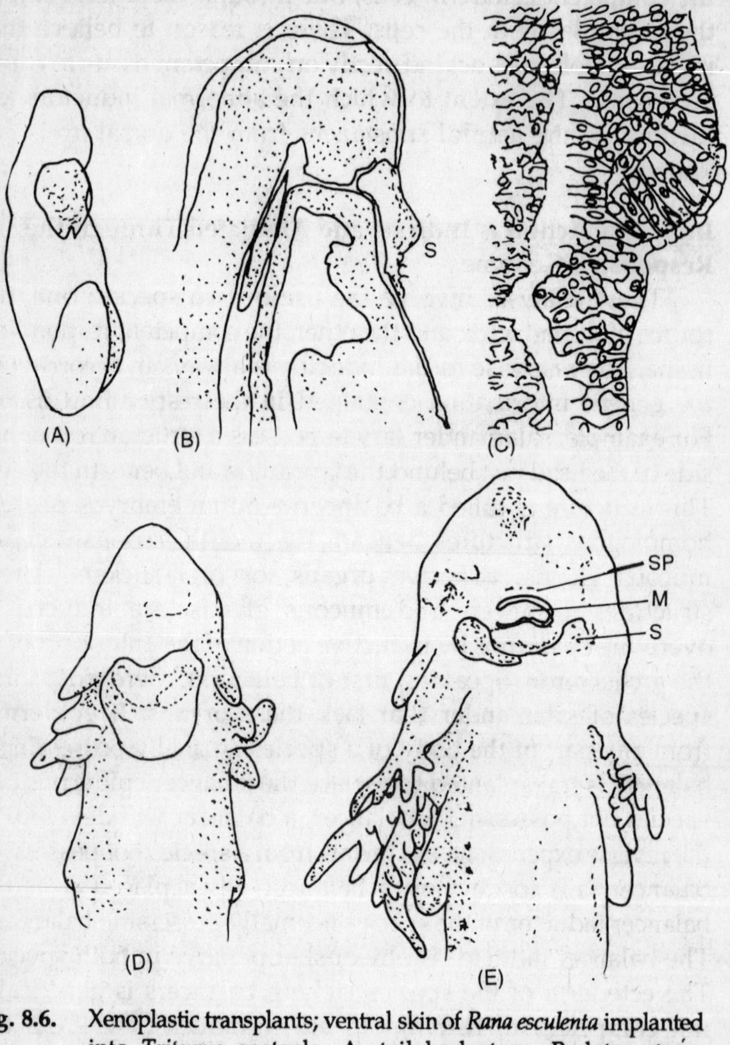

Fig. 8.6. Xenoplastic transplants; ventral skin of *Rana esculenta* implanted into *Triturus* gastrula. A, tail bud stage; B, later stage, longitudinal section; C, adhesive disc regions of the embryo of B greatly magnified; D, E urodele larva with an implant of anuran epidermis in the mouth region; E, showing sucker discs, tadpole mouth with horny jaws (M) and horny spines (SP) on the lips. S, sucker discs.

Similar experiments involve the exchange of epidermis

from the flank of frog and salamander embryos to the balancer or mucous gland region (Figure 8.6). One then has a salamander embryo (neurula) with frog belly epidermis overlying the balancer inductor and a frog embryo with salamander belly epidermis overlying the mucous gland inductor. The result is that the salamander epidermis forms a balancer on the frog in response to the mucous gland inductor of the frog, and the frog epidermis forms a mucous gland on the salamander in response to the balancer inductor of the salamander. Again, it is seen that the inductors evoke a response in the competent tissue, but the response of the competent tissue is strictly in accordance with the genomic information it possesses. The inductor itself is not a direct instructor.

CELL AND TISSUE INTERACTION IN MORPHOGENESIS

Movements of cells within the embryo make possible later inductive relationships, such as exist between chordamesoderm and overlying neural plate. Although differential growth rates, as discussed in Chapter 1, may account for some of the cell movements observed during amphibian morphogenesis, some happenings, such as gastrulation and germ layer formation, are not adequately described by models based solely on differential growth. In addition to growth patterns, some cells appear to exhibit a "preference" for certain other cells as neighbors during gastrulation and the later segregation of germ layers. Numerous experiments, including some very early ones, have been aimed at evaluating the possible elective behavior of cells during gastrulation. The most definitive experiments along these lines using gastrula material will be summarized here.

Isolation and Self-differentiation of the Endoderm

Pure endoderm isolated from the vegetal pole of an early gastrula first rounds up into spherical aggregates of cells. Neighbouring aggregates will fuse to form larger aggregates. Each sphere then flattens and the cells spread out over the

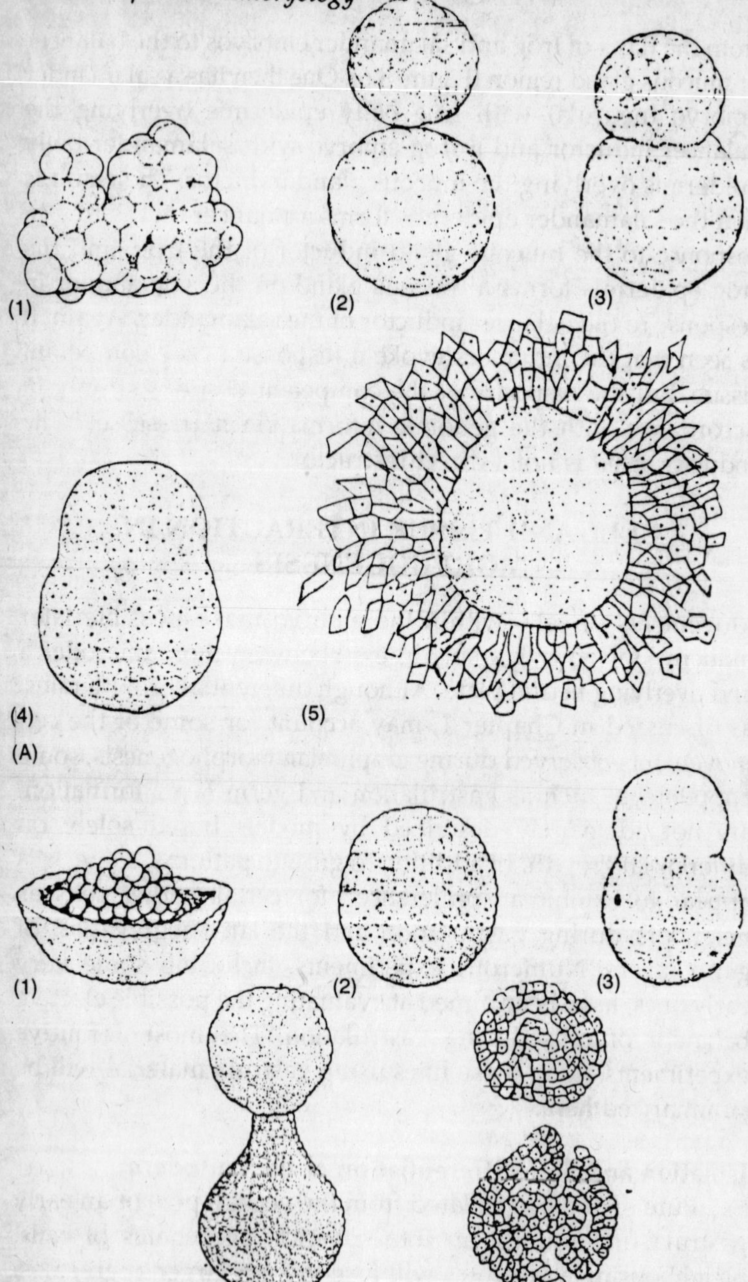
(1)
(2)
(3)
(4)
(5)
(A)
(1)
(2)
(3)
(B)
(4)
(5)

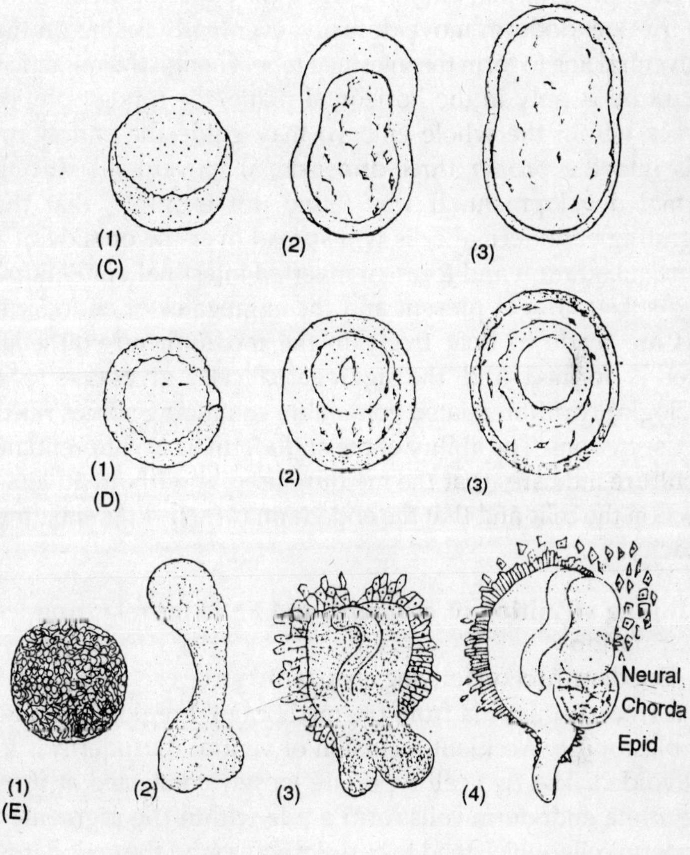

Fig. 8.7. (A) Behavior of isolated endoderm cells from the amphibian gastrula. (see text). (B) Combination of endoderm with ectoderm. (C) Combination of endoderm and ectoderm with a layer of prospective tissue interposed. (D) Endoderm surrounded by ectoderm and mesenchyme. (E) Form changes in a fragment isolated from the region of the dorsal blastopore, including ectoderm.

substratum. Mitosis is infrequent, and the cells glide over the glass surface by producing pseudopodia. A few individual cells may separate from the mass and migrate independently. The tendency to spread on the glass subtratum corresponds to the time (in a normal embryo) when the edges of the floor of the trunk endoderm move dorsally, eventually fusing on the midventral line to form the intestinal tube. Though the migration in culture is only in the horizontal plane, the topography of tissues within the whole embryo may guide the endoderm cells into the proper three-dimensional movements during normal development. It was found, for example, that the migrating endodermal cells will spread over the outside of a tissue substratum and form an inverted intestinal epithelium. If no substratum is present and the aggregates of endoderm cells are made to float freely in the medium, no epithelial sheet is formed and the aggregated cells give rise to a cytologically differentiated epithelium containing spaces filled with secretions. The ability of the endoderm to so differentiate in culture indicates that the medium used as without adverse effect on the cells and that the endoderm his self-differentiating capacity.

Changing affinition of ectoderm and Endoderm During Gastrulation

Ectoderm and underlying endoderm can be excised together from an early gastrula from the mediocaudal region opposite the blastopore. Accident inclusion of ventral mesoderm is to be avoided. The two cell types are loosely connected at first. The white endoderm cells form a pile within the pigmented ectoderm cells, which tend to curl closely around the endoderm. After one to two days the two cell types separate from one another, forming discrete spheres of ectoderm and endoderm. If a suitable substratum is present, the endoderm cells will then migrate over it is with pure endoderm isolates, and differentiate as intestinal epithelium. The ectoderm remains as a ball of epidermal cells. The separation of the two types of cells was achieved by directed autonomous cell activities and not by differential growth or budding. These kinds of separations occur between various cell types during normal development and car be thought of as arising from a negative affinity between ectoderm and endoderm. This affinity changes with time, as evidenced by the initial covering of endoderm by ectoderm, followed by the separation of the two. Isolated ectoderm three to four days old will not adhere to endoderm even when the endoderm is freshly isolated from the early gastrula.

Role of Negative Affinity in Gastrulation

The temporary incompatibility of ectoderm and endoderm seems to play an important role in steering the involution process during gastrulation. The problem of why involution rather than exogastrulation normally takes place has not as yet been clarified. In any event normal gastrulation seems to operate initially by way of an active moving apart of heterologous tissues.

Reassortment of Tissues in Explants Involving all Three Germs Layers

Endoderm and mesoderm combinations are obtained by isolating a laterocaudal piece of an early gastrula containing elements of the lateral plate mesoderm, which normally gives rise to connective tissue. Ectoderm is also included in the isolate. The isolated mass first rounds up into a sphere. The endoderm remains incompletely covered by the ectoderm. Again, after one to two days, the endoderm protrudes from the sphere and a constriction develops at the junction of the ectoderm and endoderm. Instead of progressing to complete isolation, the constriction will later relax. A vesicle is then formed with an interior of mesenchyme cells which form a loose network in a secreted fluid. The aggregates of ectoderm and endoderm cells do not remain solid, but form the walls of the vesicle. The point of juncture of the ectoderm and endoderm is marked with a slight fissure and the two are held by the internal mesenchyme. If a larger piece of ectoderm is provided, the structure resembles that diagrammed. The vesicle is completely covered by ectoderm, in two layers. The mesoderm lies between the inner ectoderm and the endoderm, which have separated a provide a cleft into which the mesenchyme can move. The inner endoderm forms a lumen lined with intestinal cells and containing a secreted fluid. The arrangement of tissues and the inward orientation of the endodermal cells in producing the intestinal epithelium are typical of the spatial arrangements found in normal embryos.

Combinations of Ectoderm and Mesoderm and Induction

When small amounts of ectoderm are isolated with dorsal lip material normally forming somites and notochord, a variety of forms results. Endoderm is excluded from these cultures. These mesodermal-ectodermal isolates first round up into spheres; then the mesoderm elongates, separates from the ectoderm, and self-differentiates in contact with the substratum into somites and notochord. The epidermis then reassociates with the mesoderm over a widening area and, subsequently, segregates into epidermis and neural tissue (through induction by the mesoderm). Thus, the affirnity of ectoderm for mesoderm progresses from a negative to a positive affinity.

The degree of specificity of cell affinities (or cell adhesion) is of the order encountered in the interaction of antigen and antibody molecules. The phenomenon is encountered in many development systems, in both plants and animals. In simple terms, it means that cells are able to "recognize" neighboring cells or neighbouring cell-bound molecules and "elect" to adhere to these or not. The phenomenon is, undoubtedly, a surface phenomenon, dependent on surface molecules present in or protruding through the surface membrane.

SELF-DIFFERENTIATION OF FIELDS AND STEPWISE DETERMINATION

An explant of presumptive mesoderm of the early gastrula is capable of self-differentiation into a variety of mesodermal tissues. The endoderm is capable, at this same time, of cytodifferentiation into intestinal epithelium. Ectoderm, destined to form either epidermis or neural tissue, is dependent upon induction by the chordamesoderm for differentiation into neural tissue. The presumptive mesodermal areas of the early gastrula are capable of self-differentiation and are referred to as fields, such as the chordamesoderm field and the heart field. If an amphibian embryo is induced to exogastrulate (using Li), the chordamesoderm will fail to underlie the presumptive neural plate area, which then develops as epidermal tissue while the chordamesoderm self-differentiates into notochord, somites, nephric structures, and so forth. The failure of the presumptive neural plate to differentiate is due to the failure of the chordamesoderm to provide the necessary inductive stimulus, as described above. If a piece of presumptive neural plate is isolated prior tc gastrulation, the result is the same. If the isolation is made after gastrulation, the explant will develop into nervous tissue. It is said to have become determined. This determination does not take place in isolated, early ectoderm.

In the embryo the entire neural axis (neural field) is determined by inductive actions along the length of the underlying chordamesoderm. In a corresponding manner, the ectoderm developing as epidermis loses its ability or competence to respond to neural inductors. As time progresses the neural axis becomes divided into subfields (forebrain, midbrain, and hindbrain; spinal cord) The regions of the neural field then act as in ductors, causing the lens to develop opposite the optic vesicle and the otocyst to form toward the posterior end of the hindbrain. The lens then becomes determined, for example, and will form lens tissue in isolation. In the intact embryo, the eye lens exerts an influence (induction) on the overlying ectoderm, causing the development of the cornea. The process of differentiation when seen in this context becomes a series of stepwise determinations and inductions, beginning with the initial action of the chordamesodermal field (determined during oogenesis) on the overlying ectoderm.

LIMB DEVELOPMENT : AN EXAMPLE OF A FIELD

The beginnings of forelimb development are to be seen in the somatic mesoderm ventrolateral to the pronephros and behind the gill region. The mesoderm becomes thickened, as does the

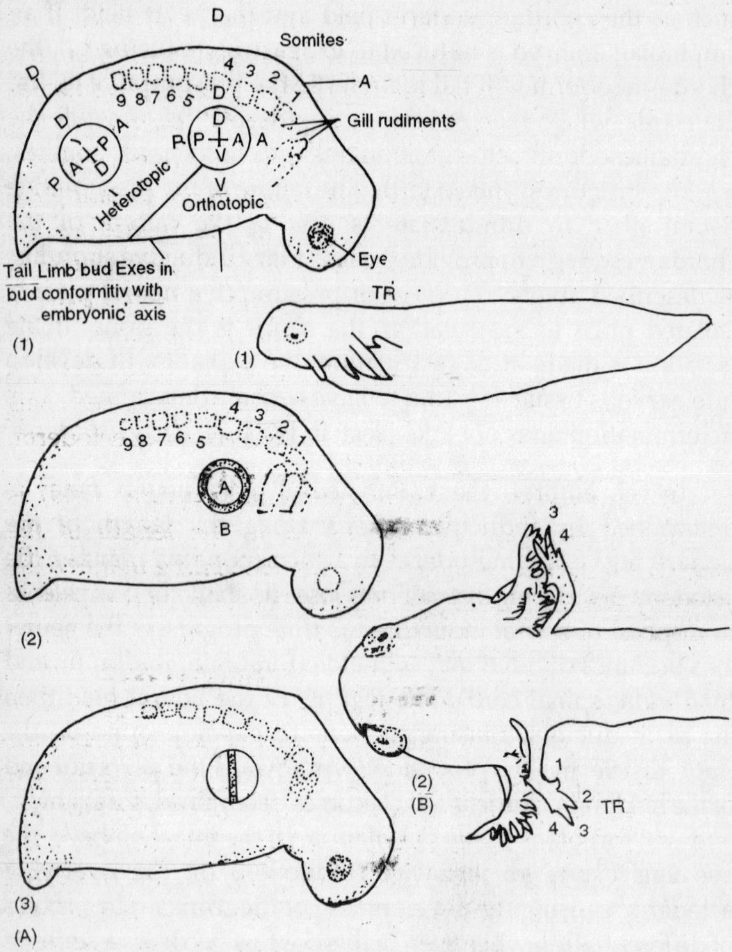

Fig. 8.8. (A) Experiments on the limb-bud differentiated in salamander (*Ambystoma*) during early tail-bud stage.

overlying epidermis. A similar development takes place just anterior to the anus. These thickenings are the forelimb and hindlimb buds, respectively. The mesoderm element in the anterior limb bud is determined at an early stage, shortly after the closure of the neural folds. Pieces of somatic mesoderm removed from the forelimb region can be transplanted to other areas of the body beneath the epidermis, where they will cause the development of heterotopic limbs (limbs forming in areas not normally giving rise to limbs). That any portion of the epidermis can, at this time, participate in limb formation is shown by experiments placing epidermis from various parts of the neurula over the limb mesoderm. All such pieces, regardless of source, are able to cooperate with the mesodermal element to produce a normal limb. This does not mean that the ectoderm of the amphibian limb bud is passive in limb development. A limb structure will not develop from isolated mesoderm; a covering of ectoderm is required even though ectoderm from a variety of sources will suffice for this purpose. The role of the ectoderm is more pronounced in amniote embryos.

The Early Limb Bud is a Regulative System

The early limb bud can be divided into two areas, or two limb buds can be placed side by side, slightly overlapping in normal orientation, or one-half of the limb bud can be removed. In all cases, a complete single limb will develop. The system is a regulative one. If the area of the limb bud is removed entirely, a peripheral area not normally involved in limb development, will replace the lost limb bud. This larger area of limb potential is the limb field. It includes the limb bud proper and a narrow zone of cells surrounding the bud which are capable of limb development, though they are not normally involved in limb formation.

Later Limb Development Involves the Stepwise Determination of the Limb Axes

The limb field is determined as limb at the time of closure of neural folds. The extent to which the limb itself is determined has been studied primarily using salamander embryos. If the early forelimb bud is moved from the right to the left side of the embryo by moving the bud around the anterior end of the embryo, the anteroposterior axis will be reversed, though the dorsoventral axis will remain in the normal orientation with respect of the embryo. This results in a limb with reversed anteroposterior polarity, indicating that this axis is determined in the embryo immediately after neurulation. At this time, the dorsoventral axis may be inverted by transplanting a limb bud from right to left by moving it over the dorsal side of the embryo, thus leaving the anteroposterior axis undisturbed. This results in a normal limb, thus indicating the ability of the dorsoventral axis to right itself, or to regulate to the pattern imposed on the limb by neighbouring tissues. In later stages, the dorsoventral axis loses this regulative ability, and inversion of this axis will result in a limb with the palmar surface facing upward instead of downward. The third axis is the proximodistal axis (in-out axis), as determined by reversal of the mesodermal element of the limb bud beneath the epidermis. It is still able to regulate at a time when the dorsoventral axis has been determined. Later stages show abnormalities of limb development if this axis is inverted, indicating a determination of this axis. These experiments are difficult to interpret precisely, since a limb cannot actually reverse itself and grow into the body; to do so would mean a loss of contact with the ectoderm essential for normal limb development, in addition to encountering structural barriers to limb growth from other body tissues.

The three major axes of the limbs are seen to be determined at separate times, and the limb bud as a whole is determined in a stepwise fashion. The later axial sequence of limb determination is in a proximodistal direction. The girdle is determined first, the phalanges last.

OTHER FIELDS

Stepwise Determination is Evident in the Development of the Central Nervous System

The normal forelimb movements of urodele larvae depend

on innervation by three pairs of spinal nerves (3, 4 and 5) comprising the branchial plexus. If this area of the neural plate is exercised and replaced with a more posterior segment, the graft will morphologically and functionally fill the role of controlling the limb movements. A similar transplant at the tail-bud stage results in limb innervation, but leads to abnormal limb movements. The posterior segment can no longer participate in the development of the branchial plexus.

At the tail bud stage, the section of neural tube which is to form the mudulla may be excised and rotated 180°. The replaced graft develops into a morphologically normal medulla, and all nervous responses of the operated larvae may be perfectly normal. Experiments on later tail-bid larvae lead to abnormalities of medulla morphology and function.

Consideration of symmetry determination in other fields (e.g., heart field) would yield similar conclusions; the field is initially regulative after being specified as a field and later becomes a determined mosaic of smaller parts, which again regulate within the subfields until they also become specified, presumably by induction from neighbouring cells. The beginning of this sequence of dependent differentiation is, again, the initial induction of the nervous system by the chordamesoderm and the continued self-differentiation of the endomesoderm.

SPECIFICATION OF SYNAPSES AND THE DEVELOPMENT OF INTEGRATED BEHAVIOUR

Background Considerations

In some invertebrates, where the total number of neurons in the animal is relatively small, the neural connections (between nerve fiber and muscle, sense organs, or other interconnecting fiber) can be mapped to some degree. Neural connections are reproduced among these animals with great fidelity, presumably on instructions from the genome. In vertebrates, there are far too many neurons for one to know with certainty the degree of order among neural connections specified by the genome.

Two extremes exist. On the one hand, neural connections might be made in random fashion. The useful connections could then be sorted out by some learning procedure. On the other hand, the entire circuitry might be specified in the genome and manifest in specific surface molecules which make each nerve cell and its connections unique. The truth most probably lies somewhere between the two extremes. Those against genetic specification sometimes argue that there are more neurons and synapses in a vertebrate brain than there are genes in the genome. However, in a number of cell recognition systems, the recognition molecules are protein-polysaccharide complexes. Specificity has been associated with the branching arrangement of the polysaccharide moiety, a property that could manifest many configurations within the instructional frame-work of a single gene. The possibility exists for complete genetic specification, though this is perhaps not the most likely situation. Alternatively, there is considerable evidence against the hypothesis of random connections later sorted out by learning activities.

The development of behaviour is obviously dependent on the development of the nervous system of receptors and effectors and interconnecting pathways (neurogenesis). Motility and bioelectric activity are essential to behavioural patterns. There is a close correlation between structural and functional maturation. Young neurons are capable of impulse transmission, and primitive synaptic connections serve to mediate early motility. Each behavioural advance follows closely on the completion of new synaptic connections. As will be seen below, sensory input and practice, or trial and error, have little to do with the development of integrated behaviour in vertebrates. In salamanders, behaviour is integrated as meaningful swimming movements from the time of hatching. In higher vertebrates such as birds and mammals, initial movements are unorganized. In the bird, these random movements are superceded by quite different, intricate, and coordinated movements that precede hatching. The latter movements do not grow out of the former, since the uncoordinated movements continue to take place even after hatching. They appear to be the result of indiscriminate and massive electrical discharges that excite many motor neurons and cause movement, which may be important for the normal differentiation of joints. The intricate hatching movements appear to be integrated from their beginning without benefit of practice movements in earlier stages. These observations tend to deny any role of sensory input in development of early integrated behaviour. It is now widely held that the neural machinery for integrated behaviour differentiated autonomously, as a "printed circuit."

Classic Experiments with Specification of Sensory Connections

These experiments deal with the regeneration of the optic nerve fibers of larval and adult salamanders and frogs. The eyes originate as a pair of lateral out-pocketings from the walls of the diencephalon. The optic vesicles reflect back on themselves, forming the two-layered optic cups. The inner layer of the cup is the retinal layer and contains the sensory elements (rods and cones), the interconnecting neurons, and the optic nerve cell bodies which send their axons to the optic lobes (optic tectum) in the midbrain region. Experiments with the optic tectum in the midbrain have established that the retinal field is projected, point by point, onto the optic lobes. Specific stimulation of the retinal cells (one or two cells at a time) gives signals to specific cells of the lobe corresponding to the portion of the retinal field being stimulated. The optic fibers are predetermined as to their point of termination in the optic lobe even before they have grown back to the midbrain. This can be shown by rotation of the optic cup prior to optic nerve growth. This operation results in inverted vision, leading to the conclusion that the optic nerve fibers deviated from their normal path to journey from their position in the inverted retinal layer to their normal terminals in the tectum.

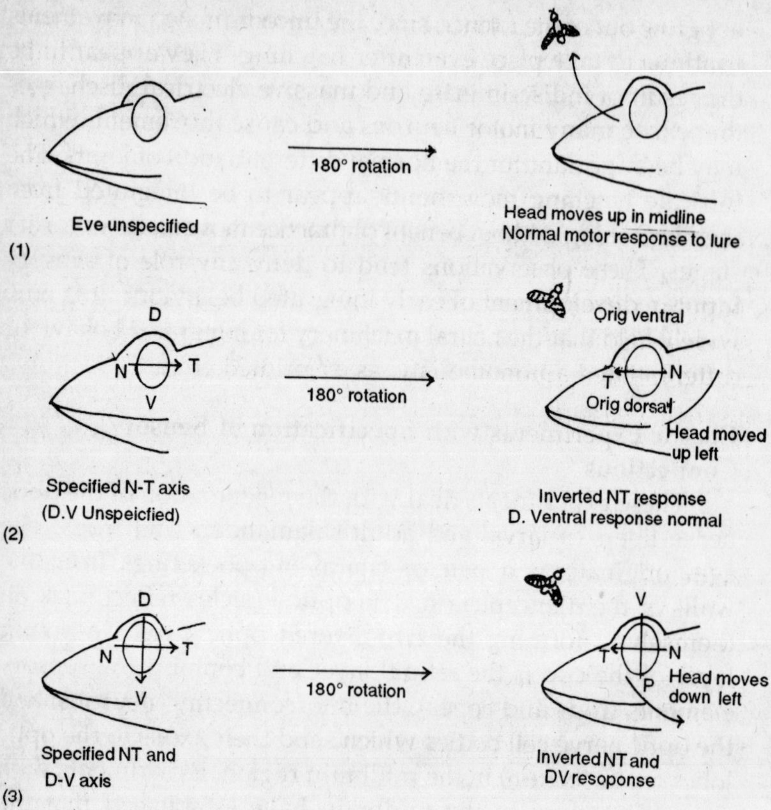

Fig. 8.9. Diagram of development of retinal polarity in eyes of amphibians rotated at various embryonic stages. Embryonic state of determination at rotation in left column; resulting adult behaviour in right column. DV, dorsoventral; NT, nasotemporal. The fly acts as a lure.

Crushing of the optic nerve in salamander larvae or adult frogs can result in optic nerve regeneration and restoration of normal vision and normal responses to visual cues in the environment. Rotation of the eye in the orbit after crushing the optic nerve also results in the restoration of vision in a significant number of operated animals. The vision of the inverted eye is, however, inverted, as are the responses to visual cues in the environment. The implication of these studies is that the optic nerve fibers returned to their original connections in the optic lobes through a thick mat of apparently randomly regenerating nerve fibers. Now the environmental stimulations impinge on inappropriate portions of the retina (a dorsal stimulation now focuses on the ventral retinal field, for example). Responses to the stimuli are correspondingly inverted. The inappropriate responses continue throughout the life of the adult frog. Most animals continue to snap at lures or food (flies) in a reverse manner; a fly dangled dorsal to the nose will be lunged for between the front legs. These responses in some animals were replaced by indifferent reaction, as though the frog had simply given up on the impossible chase. Thus, there may be a tendency in some to ignore the lure or to inhibit the responses. However, in no animal was a positive attempt at correction of the defect noted.

Evidence from Lesions in the Optic Tectum of the Optic Lobes

Specific lesions (scotomas, or blind spots) produced in the tectum result in production of expected visual deficiencies in animals with regenerated optic nerves. An anterior lesion in the tectum abolished responses in the anterior region of the visual field, and so with lateral and posterior lesions. In those animals with inverted eyes, the lesion in the posterior region produced deficient responses in the anterior visual field. In summary, it can be concluded that the ingrowing optic fibers reestablish functional connections in the same topological areas of the optic lobe where they originally terminated during normal neurogenesis.

Implications of These Studies

The foregoing data support the contention that each retinal locus possesses functional connections with brain centers differing from all other loci. The data also support the contention that the orderly restoration of central reflex connections, which occur regardless of the orientation of the retina, must be independent of functional adaption or learning. It is also implied that the retinal fibers must possess specific properties by which they are recognized in the optic lobe according to their respective retinal origins. In studies on regeneration of the optic nerve in fish, these conclusions have been borne out by direct anatomical observation, not only with respect to the terminal connections established in the tectum, but also with respect to the pathways taken by the optic fibers to reach their central, terminal points.

Experimental Evidence; Motor Activity

The concept of the "printed circuit" of neural activity as a part of the genetic endowment of a species was supported by early narcotization experiments on salamander embryos. In this study a clutch of salamander eggs was divided into two groups. One group was kept as a control. A second group was anesthetized with chloretone from the preswimming stages to beyond the swimming stage. When the anesthetized tadpoles were removed from the anesthetic, they were found to swim in a fashion equalling the coordination of the control group.

Additional Information from Inverted Limbs

The patterning of the central neural pathways would thus seem to depend on specific neuronal connections. This type of selective connection (or affinity) is presumably based on molecular recognition of one nerve cell by another, or of recognition of particular muscles by nerves. Evidence for a chemoaffinity (as opposed to trial and error) mechanism operating between nerve fibers and between nerve and muscle is found in studies of inverted salamander limb buds. These limb buds were inverted at a time when their anteroposterior axis had been determined, and thus grew out in a reverse position. It was found that from the beginning of motility the limbs also moved in reverse. This implies that the spinal center for locomotion develops independently of the inverted limb and produces a "malfunction" that is never corrected. In another experiment, the legs were deprived of sensory nerves before leg movement had started. Coordinated locomotor function was not impaired in the treated limbs. It has also been found that the nerves of the lumbosacral plexus (8, 9, 10) will grow around a mica barrier, taking an abnormal path to the developing hindlimb. The normal innervation of the hindlimb is possible if the barrier is not too extensive. This points to a selective affinity of these nerves for muscle fibers of the hindlimb.

It has been found that a forelimb transplanted to the head region will be innervated by cranial nerve fibers, sensory and motor. The limb will not function in coordinated fashion with other limbs, but rather tends to move in concert with anterior structures (e.g., movements of the gill chamber) also innervated by the nerves supplying the abnormally placed limb. The abnormally placed limb is innervated non-specifically in an abnormal situation. Under these conditions it functions as though it were a part of the motor apparatus in the abnormal position and not in co-ordination with other normally placed limbs. The basis of the apparently non-specific attraction of cranial nerves to the abnormally placed muscle is puzzling and represents a question to be resolved.

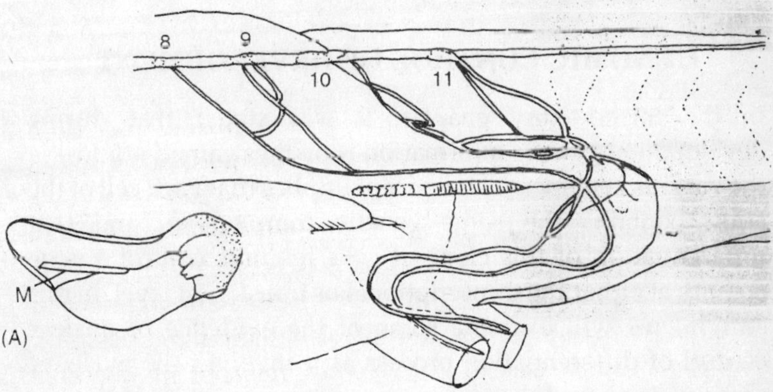

Fig. 8.10. Response of outgrowing nerve fibers to physical barrier in normal pathway. (A) A mica plate (M) inserted in the path of outgrowth of the spinal nerves to the hindlimbs in a frog embryo: (B) the nerves have grown around the obstacle and have reached the hindlimbs.

Gradual Specification of the Nervous System

Beginning with the induction of the central nervous system by the dorsal lib of the blastopore, the development of the functional nervous system proceeds through a period of plasticity to an extremely high degree of specification. Nerves maintain the same specific connections made during development after regeneration, so that motor nerves innervate the appropriate muscle, and sensory nerves innervate the proper end organ in establishing reconnections. Clinical attempts in humans to supply facial muscles with cervical nerves (as when the facial nerve is paralyzed) may result in innervation of facial muscles by the cervical nerve. However, the facial muscles function in concert with cervical voluntary muscle rather than with other facial muscles. This situation is not corrected centrally and persists throughout life. In the section on human development additional information is available on plasticity during development and on the essentially inflexible situation found in the adult. The basic mechanism governing nerve cell outgrowth and synaptic connections is essentially unknown and represents one of the most fascinating and challenging areas of experimental biology.

GENOMIC CONTROL OF DEVELOPMENT

In the introductory chapter, it was stated that during development genetic information is neither gained nor lost in the various cell lines. This concept implies that each cell of the adult contains the same genetic complement, and that differentiation of this clone of cells requires control a gene activity at either the transcriptional or translation level. In this section, we will examine some of the evidence of nuclear control of differentiation process as it involves the nuclei of embryonic and adult cells.

Cleavage Nuclei are Equivalent

Cleavage nuclei were first tested by a simple but ingenious experiment. The fertilized salamander egg was constricted by a hair loop into a dumbbell shape. The fusion nucleus was located in one portion that proceeded to cleave, while the portion without the nucleus remained unsegmented. The constriction is made parallel to the anterior-posterior axis of the embryo and bisects the gray crescent. After cleavage to perhaps 32-64 nuclei, the constriction allows one of the cleavage nuclei to pass to the enucleated egg. The selection of the nucleus which passes to the undivided side is completely random. The newly nucleated portion of the egg then begins to segment, and the loop is pulled tight to completely separate the two halves. The result is the formation of two identical twins, one of which will be somewhat younger as regards the stage of development. Since the selection among the cleavage nuclei was completely at random, it follows that all cleavage nuclei are equivalent and totipotent at the stage at which the nucleation of the uncleaved portion of the egg took place.

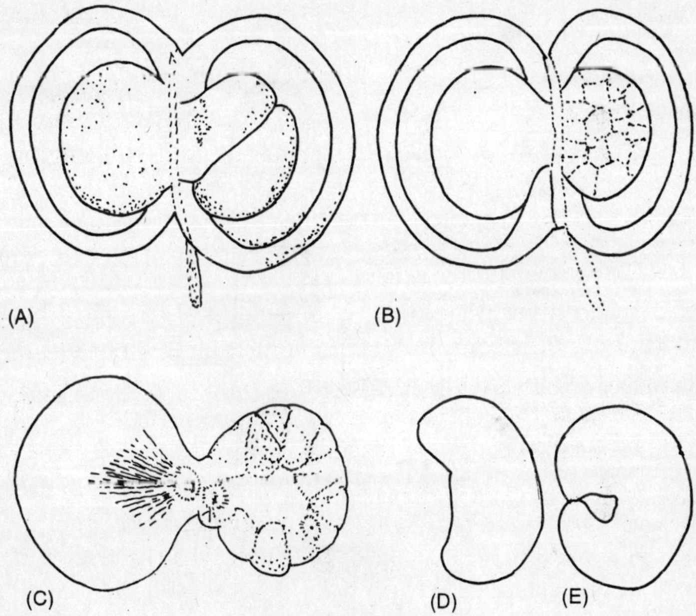

Fig. 8.11. Nuclear equivalence during cleavage. A restrictive noose is placed around the egg at the first cleavage (A, B). The nucleate half cleaves. Later (C) a cleavage nucleus passes into a nucleate half. resulting in older and younger twins.

Equivalence is Extended to the Blastula Through Nuclear Transplantation

In the 1950s the technique of nuclear transplantation was

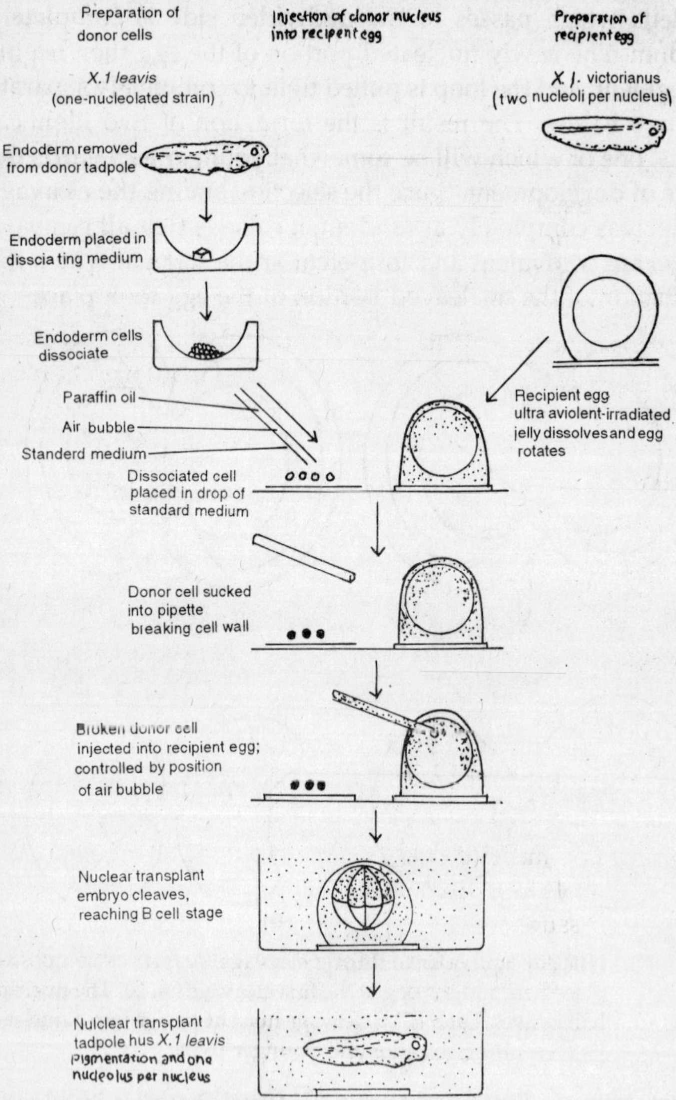

Fig. 8.12. Nuclear testing in *Xenopus* by nuclear transplantation.

developed using the frog egg. This procedure allows for the direct testing of embryonic nuclei for equivalence and totipotence. In essence, the technique involves the enucleation by microsurgery of an artificially activated egg by removal of the second polar spindle and associated chromosomes. This is followed by the introduction of a test nucleus into the egg via a micropipet. The test nucleus is taken from any source by pulling a cell into a small-bore pipet with gentle suction. This procedure usually breaks the cell membrane. Transplantation of isolated nuclei without transfer of cytoplasm has not been accomplished, since nuclei isolated in artificial media are inactivated as far as future nuclear division are concerned. The transplanted cytoplasm is diluted by the egg cytoplasm by about 1: 40,000 and is usually neglected in interpretation of results. By the late blastula stage, the original nucleus has proceeded through 13 or 14 generation producing about 5,000-16,000 cells. About 80 per cent of the eggs receiving transplants of a blastula nucleus cleave, and a smaller proportion proceed through all stage of embryonic development and metamorphosis. This experiment directly demonstrates the equivalence and totipotence of nuclei of the late blastula. The technique has been extended to the gastrula and later stages, nuclei of cells of an adenocarcinoma of the adult frog kidney.

Transplants from Later Stages Prove Difficult to Interpret

Using *Rana pipiens,* the percentage of transplant embryos showing normal development is markedly reduced as the age of the donor nuclei increases. However, in *Xenopus,* gastrula nuclei, tadpole intestinal nuclei, and nuclei from the germ line and adult tissues (excluding brain nuclei) reportedly give rise to normal tadpoles, as do the nuclei of the frog adenocarcinoma. Part of the difficulty with nuclei from embryos of later stages may be technical and involve difficulties in handling the small cells without injury to the nuclei and chromosomes. There is also a biological difficulty resulting from the faster generation times during cleavage as compared to later stages (about once every hour or two versus once every two or more days). The replication of chromosomes may fail to keep up in transplanted nuclei from older embryos or adults, leading to partial replication of chromosomes may fail to keep up in transplanted nuclei from older embryos or adults, leading to partial replication of chromosomes and loss of chromosome material. Fragmentation of chromosomes is observed in *Rana* nuclei transplanted from later stages.

Success obtained with older material might be due to parthenogenesis, resulting from failure of the experimenter to remove the egg nucleus. Such operational errors have been recognized and eliminated by using markers in the transplanted nuclei. Both triploidy and number of nucleoli (Figure 8.13) have been used as cytological markers for transplanted nuclei. Genetic markers could be used as well. The marker appears in all the cells of the embryo resulting from nuclear transplantation if the transplant nucleus has indeed participated in cleavage in the absence of the egg nucleus. The successful nuclear transplants in the *Xenopus* system, as well as those using adenocarcinoma nuclei of frog kidney, utilized marked nuclei. There is no mistake about these positive results. It has been argued, however, that the percentage is so low as to raise the question of the involvement of exceptional, undifferentiated cells in the successful experiments.

The Basic Question Involves two Concepts

The first is the question of the totipotence or pluripotence of the somatic cell nucleus. The second is the question of the reversibility of nuclear changes (differentiation) which occur during development. Somatic nuclei may be possessed of all their nuclear genome and may thus potentially function as zygote nuclei. However, the process of differentiation, involving selective inactivation of most of the genome, may require special circumstances for reversal which are not met by the activated egg cytoplasm. Indeed, the nuclear changes may not be reversible at all. This is certainly the case in somatic tissues of some insects which develop giant polytene chromosomes. In these cells the entire genome is represented, but each chromosome has been replicated throughout its length perhaps 1000-fold, leading to structural specializations which preclude its function as zygote nucleus, perhaps on purely mechanical grounds. Systems other than nuclear transplantation (such as the development of tumors of primordial germ cells) also indicate totipotence of nuclei from differentiated tissues. It is necessary to consider all the evidence at hand and not just nuclear transplantation data alone, in arriving at the conclusion that the genome is conserved during development.

The Cytoplasm Controls Nuclear Activity

At the molecular level, the transplanted nuclei undergo changes similar to those noted in the male pronucleus after normal fertilization. The sperm nucleus first swells and then begins synthesis of DNA. A transplanted nucleus from the midblastula stage, normally synthesizing DNA, will increase in volume and continue DNA synthesis at the times specified by the egg cytoplasm. Neurula nuclei, normally synthesizing ribosomal RNA but infrequently DNA, will, upon transplant to an activated enucleated egg, cease ribosomal RNA synthesis and begin DNA synthesis after the nuclear-volume increase. During cleavage, these nuclei synthesize some messenger RNA. The transplanted neurula nuclei will again synthesize ribosomal RNA upon reaching the gastrula and neurual stages, following the transplant. Nuclear transplants may also be made to primary oocytes (actively synthesizing RNA, not DNA) and to unfertilized, unactivated eggs (making neither DNA no RNA). In growing primary oocytes, transplanted nuclei from blastulae are induced to cease DNA synthesis and commence RNA synthesis. In unactivated eggs, nuclei are quiescent with respect to RNA synthesis. The transplant chromosomes condense to resemble the egg meiotic chromosomes. Foreign nuclei (mouse liver) transplanted to enucleate, activated frog eggs are induced to synthesize DNA and to stop RNA synthesis. Clearly, changes in nuclear activity are dictated by the cytoplasm, and in a universal language as well. The several responses of nuclei to different cytoplasms argues for pluripotence of these nuclei.

Preprogramming of the Oocyte and RNA Synthesis During Development

Transcription of the genome occurs during oogenesis, as well as during embryogenesis. The question arises as to the qualitative nature of the mRNA formed at various times between oogenesis and the completion of differentiation. Some mRNA is transcribed during oogenesis for later use by early embryos (through blastula). To this extent, the egg may be said to be preprogrammed with differentiation controlled at the translational level. The paragraphs below examine gene activity during oogenesis through late tadpole stages, with respect to synthesis of the various classes of RNA (ribosomal, transfer and messenger).

During oogenesis lampbrush chromosomes are present, as are numerous nucleoli. The nucleoli have been found to represent amplified DNA sequences engaged in the production of ribosomal subunits. This gene amplification represents a specialization for the accumulation of the large amount of ribosomal RNA required during early development through to the hatching tadpole stage. The rate of ribosomal RNA synthesis is maximal at the lampbrush stage and predominates to the point of obscuring synthesis of messenger RNA. In mature oocytes, 95 per cent of the total RNA is ribosomal. Though measurement of the template mRNA is difficult in the presence of large amounts of ribosomal RNA, it has been estimated that about 2.5 per cent of the total RNA of *Xenopus* oocytes is template active. This template or messenger RNA seems to have been synthesized during the diplotene lampbrush stages, particularly stage 4, and constitutes the bulk of the messenger RNA inherited by the embryo as maternal template. In terns of unique-sequence DNA transcripts oocyte RNA contains about 9 per cent of the information content of unique-sequence DNA. This is remarkable close to the value for *Urechis* where a similar complexity is found. At ovulation, a burst of messenger RNA synthesis occurs under the influence of pituitary hormones. It is of some interest that the egg cytoplasm does

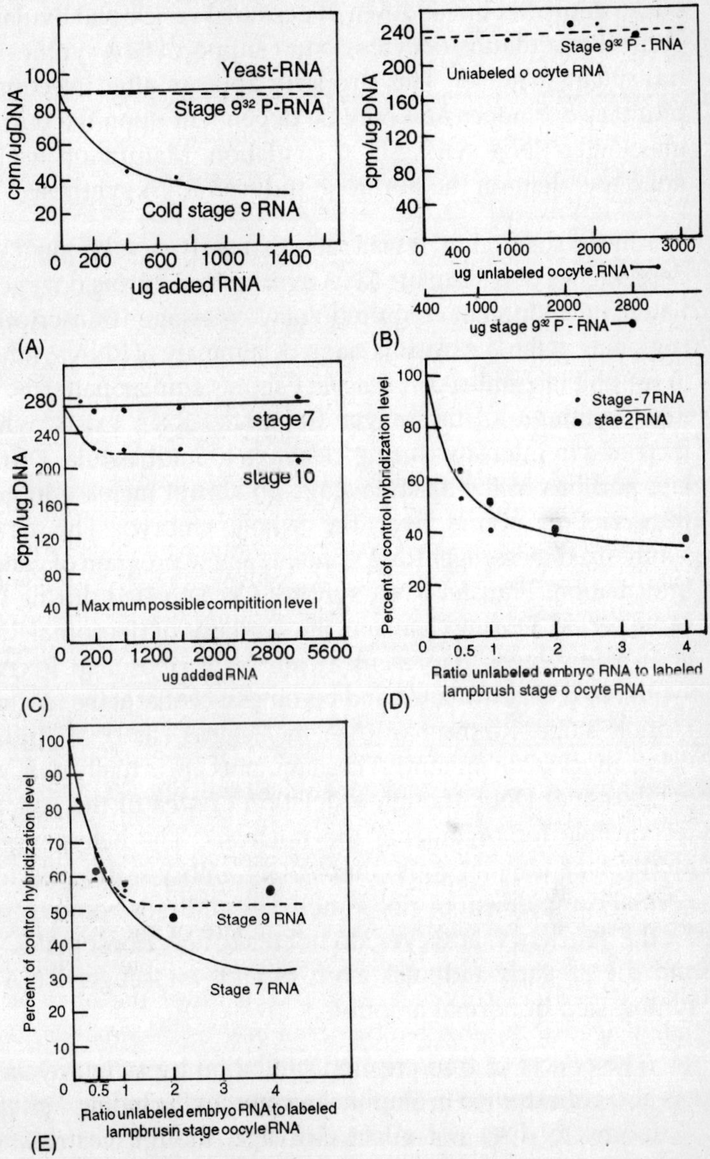

Fig. 8.13. (A) Competitive DNA: RNA binding experiments indicating degree of homology between RNAs of various embryonic stages and RNAs of lampbrush oocytes of *Xenopus*.

not acquire the ability to induce DNA synthesis in transplanted nuclei until after break-down of germinal vesicle and ovulation occur. Preovulation oocytes do not support DNA synthesis in transplanted nuclei. This property appears after injection of pituitary hormones and may be dependent upon the burst of messenger RNA synthesis at ovulation. Maturation itself is not dependent on the hormone-induced RNA synthesis.

In the sea urchin, it was seen that early development was dependent upon template RNA synthesized during oogenesis. Later development required new message transcription, beginning at the late blastula stage. A summary of RNA synthesis in amphibian embryos (*Xenopus*) shows similar patterns. An early synthesis of messenger (template) RNA exists, which increases in intensity during cleavage to midblastula. During late portions of the blastula stage an abrupt increase in gene transcription occurs over the whole embryo. The newly synthesized messenger RNA contains a new program of genetic information. Transfer RNA synthesis is activated during this same period. During cleavage, the synthesis of ribosomal RNA is at a low level, if present at all. New ribosomal RNA is synthesized at gastrulation and becomes essential at the hatching tadpole stage. Ribosomal RNA in *Xenopus* has been studied using anucleolate mutants. These mutants are actually deficient in ribosomal DNA though a deletion of area of the genome responsible for organizing the nucleous. The heterozygous mothers of homozygous anucleolate embryos synthesize a normal complement of ribosomes. The embryos homozygous for the deficiency, however, do not make new ribosomal RNA and die as early tadpoles even though messenger RNA is synthesized in normal amounts.

The effects of transcription inhibition by actinomycin D has also been studied in amphibian embryos, including *Xenopus*. Actinomycin does not affect cleavage, though gastrulation and neurulation—are completely blocked. Inhibition of translation using puromycin blocks cleavage in *Xenopus*, as it does in the sea urchine embryo.

Organization of the Xenopus Genome

Studies involving reannealing of sheared *Xenopus* DNA fragments indicate interspersion of longer section of single copy DNA with shorter, repetitive segments. The sheared fragments of various lengths were incubated to a Cot value to include only repetitive segments, using 450 nucleotide long DNA fragments in excess to drive the reaction. As the length of the fragments increased, these longer fragments contained more and more unique-sequence DNA, a result expected if the single copy sequences were interspersed with repetitive sequences.

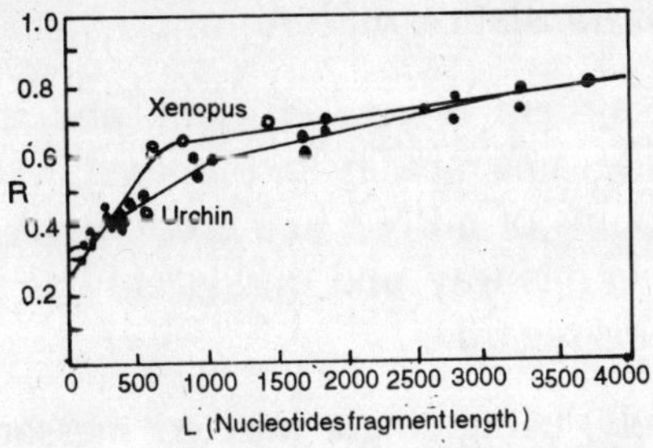

Fig. 8.14. Interspersion of single-copy and repetitive DNA in *Xenopus*. Relation between fragment length and hydroxyapatite binding in the DNA of the sea urchin (*Strongylocentrotus purpuratus*), compared to similar curve for Xenopus DNA.

The use of Xenopus Oocytes in mRNA Assay; the Question of Translational Control

Oocytes contain mRNA which is transcribed during oogenesis and "masked" for controlled use during early development. The nature of the mask is not known. However, mRNAs ("unmasked") from various diverse sources can be microinjected into oocytes and will be translated along with the messages normally available for protein synthesis in these cells. The oocytes are readily permeable to amino acids, and the newly synthesized proteins can be labeled and separated from one another by various chromatographic or immunological procedures. A number of species of eucaryotic mRNA have been used, in purified form, to test the oocyte system. Included are mRNAs for calf lens protein, duck, rabbit and mouse globin chains, trout protamine and rabbit hemoglobin. To reemphasize, the protein products of translation of mRNAs are stable and have been characterized as belonging to the species providing the mRNAs and not to the frog. Since oocyte is not engaged in synthesis of any of the proteins coded by the exogenous mRNAs, recognition of the translation product is possible. The fidelity of translation has been examined and found to be of the same order as when the mRNA is translated in the donor cells. This reflects a marked lack of species specificity in the process of translation mRNA.

The oocyte system is very efficient, and messages are translated with the same rapidity encountered in normal donor cells. Small amounts of mRNA in a crude RNA preparation can be detected in this way, and the system has proved to be most useful in this regard.

Quantitatively, using single injected messages, a linear relationship between mRNA injected and protein synthesized exists up to a saturation level. Even at saturation the amount of endogenous oocyte protein synthesized is not altered. This would indicate that the oocyte has a definite reserve translation capacity beyond normal endogenous levels, which may not be the same for all mRNAs injected. Some have argued that the presence of this spare translation capacity rules against controls at the translational level. To put it another way, it appears that any mRNA present in the cytoplasm will be translated, regardless of origin. The translation activity of exogenous message continues for days in injected oocytes. Normal translation activity is controlled by "masking" endogenous mRNA. The proper masking control for exogenous message seems not to exist. Translation can proceed in embryos in the absence of transcription (actinomycin studies) in controlled fashion, and one must conclude that the "foreign" message is somehow treated differently from that produced by the oocyte nucleus. A number of controls ("masks") have been proposed. It is possible that oocyte messages are transcribed, but are retained in the nucleus until time for translation. It is also conceivable that messages are produced by the oocyte nucleus and immediately combined with a cytoplasmic repressor that is removed at the proper time, such as after fertilization. These controlling events may be inoperative in the injection of foreign messages into the oocyte cytoplasm.

Both transcriptional and post-transcriptional controls are utilized by cells. A major problem for developmental biologists is evaluation of the relative importance of these levels of control in bringing about and maintaining a differentiated state. Repressing a gene at the level of transcription of DNA might seem more economical than synthesis of message to be followed by the masking or inactivation of that message. Transcriptional control is seen in terminal stages of development, in which all genes save one sequence may be inactive. In early development some control seems to be at the level of controlling translation of preexisting maternal message, as occurs in sea urchins and amphibians.

Between the two extremes, the cell probably utilizes both methods. Under certain conditions, transcription may be speeded and translation of mRNA accelerated, either by increased translational rate, or by decreased degradation of the message. Controls are therefore complex and involve levels of controls not only at the DNA and at the ribosome, but also in the length of functional life of the mRNA molecules. Most differentiated tissues do develop very stable mRNA molecules for production of tissue-specific proteins. The cells of the eye lens, as epithelial cells, synthesize the crystalline protein of the lens fibers. Actinomycin inhibits this synthesis in the lens epithelial cells, but not in the fiber cells into which the epithelial cell develop. The fiber cells possess an extremely long-lived mRN that operates in the absence of any transcription for long periods of time, obviating the need for nuclear activity. A similar situation exists in cells synthesizing hemoglobin. In this respect, translation control and long-lived messages seem to be characteristic of the beginning and the end of development.

Transcriptional control seems to operate during differentiation, though information in support of such generalizations awaits a thorough analysis of at least one example of cell differentiation.

Sources of Information and Raw Materials for Early Protein Synthesis

Most of the evidence concerning the function and fate of early messenger RNA tends to support the suggestion that these molecules are largely stored for later use in protein synthesis. The observation that both enucleate and actinomycin-treated embryos synthesize as much protein as control embryos during cleavage suggests that the amount of protein synthesis dependent upon newly synthesized messenger RNA is, at least quantitatively, insignificant. Protein synthesis, as measured by incorporation of labeled amino acids, takes place in the unfertilized egg. No increase in protein synthesis occurs after fertilization. What is actually being measured by following incorporation of radioactive amino acids into protein is the synthesis of new protein molecules from a pool of amino acid precursors derived from the breakdown of yolk platelets. The yolk protein is thus broken down and resynthesized into different protein molecules according to the needs of the embryo. Some protein molecules may be qualitatively the same as those made in the oocyte. New types of protein appear at gastrulation, as evidenced by serological experiments. These experiments demonstrate that antigens are present in the gastrula and neurula which are absent from the earlier stages. These proteins may be translated from the messages synthesized during the burst of messenger RNA synthesis that occurs just prior to gastrulation. They may also be the result of the unmasking and translation of oocyte mRNA already present. In any event, the rate of protein synthesis does increases during gastrulation.

9

Metamorphosis in Amphibia

A remarkable series of changes takes place when an aquatic, fishlike tadpole is transformed into a land-dwelling frog. Throughout embryonic and larval life the development of the frog has been almost entirely a gradual, unfolding process of growth and differentiation, but during the metamorphic period growth and differentiation, but during the metamorphic period occur simultaneously with dramatic suddenness. The term "metamorphosis" has broad zoological usage. It is used to designate any group of developmental changes that are completed within a period of time which is brief in comparison with the full developmental period of the individual concerned. A variety of animals undergo metamorphoses of diverse sorts. Each of the events of amphibian metamorphosis contributes toward the preparation of the developing individual for life on land, but one cannot say that preparation is confined exclusively to the metamorphoic period, since, after all, the limbs and the lungs have been differentiating during most of the larval period.

Staging of Anuran Development

Development is, of course, a continuous process, but in order to make critical comparisons between different species or between different individuals of the same species it is desirable that the developmental period be divided into discrete steps or stages and that they be described in a staging table. A number of such tables have been prepared for different amphibians. The ones for anurans with aquatic larvae are quite similar to one another. Gosner (1960) has combined two of the best of them — those of Limbaugh and Volpe (1957) and Taylor and Kollros (1946) — into a simplified and generalized series that should be readily adapted for use with any frog or toad having free-swimming larvae. The larval and metamorphic stages of Gosner's series are presented here because they illustrate the sequence of external changes that are most easily observed during typical anuran metamorphosis.

That stages 23, 24 and 25 are marked by the formation of the operculum and the consequent covering over of the external gills by its anterior component which grows backward from the hyoid arch. By stage 25 the anterior and posterior components of the operculum fuse except for a small funnel-shaped opening to the outside, the spiracle.

At about stage 23 the oral disc and rows of keratinized labial teeth and horny beaks (Fig. 9.3begin to form. In stages 23 to 25 the initial formation of pigmentary patterns takes place, chromatophores of several types appearing at about stages 23 and 24.

In distinguishing between families the diagnostic feature is the form of the oral disc. Its essential peculiarities are present by about stage 26, although changes occur subsequently in the number and form of oral papillae. The tooth rows develop gradually. Although the "mature" tooth row formula of a species is usually established in the early larval stages, the relative proportions and numbers of the rows sometimes change during ontogeny.

Independent feeding commences sometime between stages 25 and 26. Primarily on this basis stage 26 is arbitrarily designated as the first larval stage.

Identification of stages 26 to 40 is made by examination of the hind limbs. Stages 26 to 30 are distinguished by changes in the ratio of length to diameter in the hind limb primordium. At stage 31 the "foot" is paddled-shaped, and in subsequent stages the configuration of the paddle is altered as individual toes are differentiated. The distinguishing criteria of stages 23 to 40 are proportional changes in the length of individual toes and in the appearance of metatarsal and subarticular tubercles.

Etkin (1932, 1955 and 1964) regards anuran metamorphosis as consisting of two phases, prometamorphosis and metamorphic climax. Prometamorphosis begins at stage 36 and is marked by the rapid elongation of the hind limbs. During stage 40 the cloacal tail piece is resorbed, shifting the position of the anus. The completion of this shift marks the beginning of stage 41. The operculum thins and becomes translucent over the developing forelimbs, forming a "skin window" through which the forelimbs erupt.

The appearance of the forelimbs marks the beginning of stage 42, the first stage of metamorphic climax. Stages 42 to 46 are characterized by the remodeling of the head and are distinguished primarily by changes in the mouth. Larval mouthparts are lost. The tympanum is formed. At stage 46 metamorphosis is essentially complete. Newly transformed young may or may not resemble the adults sufficiently to permit positive identification.

The extensive alterations in external features noted are accompanied by equally profound changes in internal anatomy such as the adaptation of the aortic arches and the hyoid cartilages and muscles for pulmonary respiration, the loss of the lateral line system, the development of a muscular tongue, and the shortening of the gut. In this chapter, however, we shall give primary attention to external morphology.

Typical Development Among Urodeles

In the urodele the events of metamorphosis appear to be much less striking than in the anuran since the superficial resemblance of the salamander larva to the adult is much greater than that of the tadpole to the frog.

Rather abbreviated sequences of metamorphic stages have been described for *Eurycea bislineata* by Wilder (1925), for

Ambystoma maculatum, Notophthalmus viridescens, A. jeffersonianum* by Grant (1930a and 1930b), and for *Gyrinphilus palleucus* by Dent and Kirby-Smith (1963) but external morphological changes are not sufficiently marked to substantiate a long series of stages such as that described for anurans. The only truly obvious change that is a consistent feature of metamorphosis among all urodeles is the resorption of the gills (Figs. 9.2, 9.3 and 9.4). Other consistent but less obvious external features are the resorption of the tail fin and the larval folds of the lower jaw, fusion of the gill slits, and the differentiation of eyelids. Also, colours and colour patterns typical of certain species change from larval to adult shades and configurations. In addition to the events common to all urodeles, there are numerous changes in features peculiar to certain groups. For example, the Plethodontidae usually develop naso-labial folds during the latter states of metamorphosis.

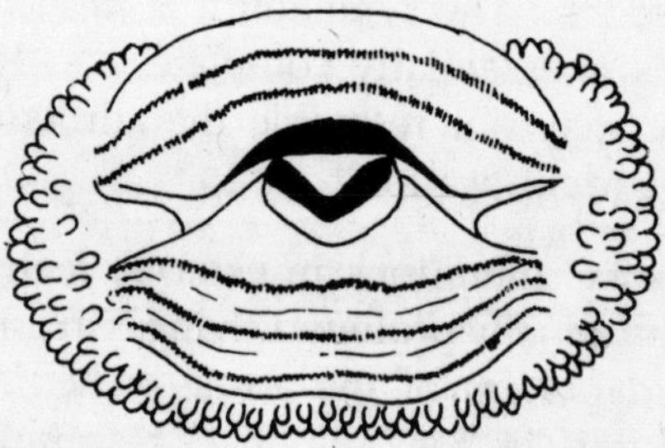

Fig. 9.1. Detailed diagram of oral disc from tadpole of *Leptodactylus melanotus*, a typical aquatic anuran larva. On the lateral and ventral borders are oral papillae.

The internal changes of urodele metamorphosis are most extensive in the head region. Valves develop around the nares to prevent the entrance of water if the adult becomes submerged.

* Although *Notophthalmus* is the generic name currently accepted by systematists for the eastern spotted newt, this animal has at different times also been called *Diemictylus* and *Triturus*. It became well established as subject for experimental studies during the time it was known as *Triturus*, and many continue to refer to it by that name in the literature of experimental zoology.

Alterations take place in the skull and in the hyobranchial apparatus. The skeletal changes are accompanied by alterations in associated musculature. In the integument there is a loss of large glandlike cells, the Leydig cells, and the formation of a keratinized stratum corneum.

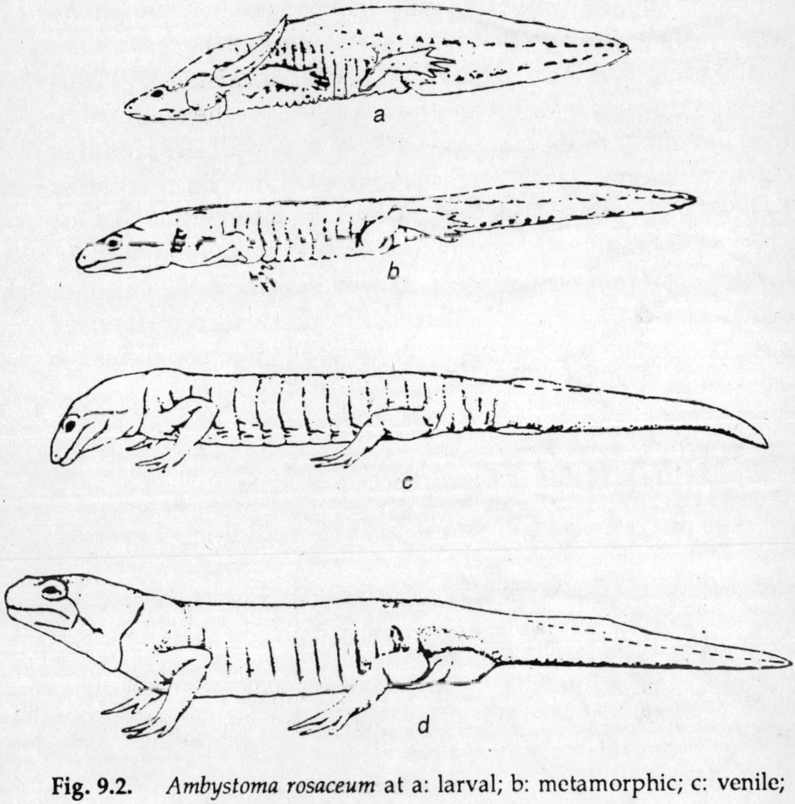

Fig. 9.2. *Ambystoma rosaceum* at a: larval; b: metamorphic; c: venile; and d: adult stages.

Variation in Amphibian Life Histories

Although there are many typical anurans and urodeles whose life histories follow the patterns that have just been described, one does not venture far into amphibian systematics and biology without discovering that there are wide variations upon the basic theme. In the evolution of the vertebrates the amphibians provided the link between the fishes and the reptiles.

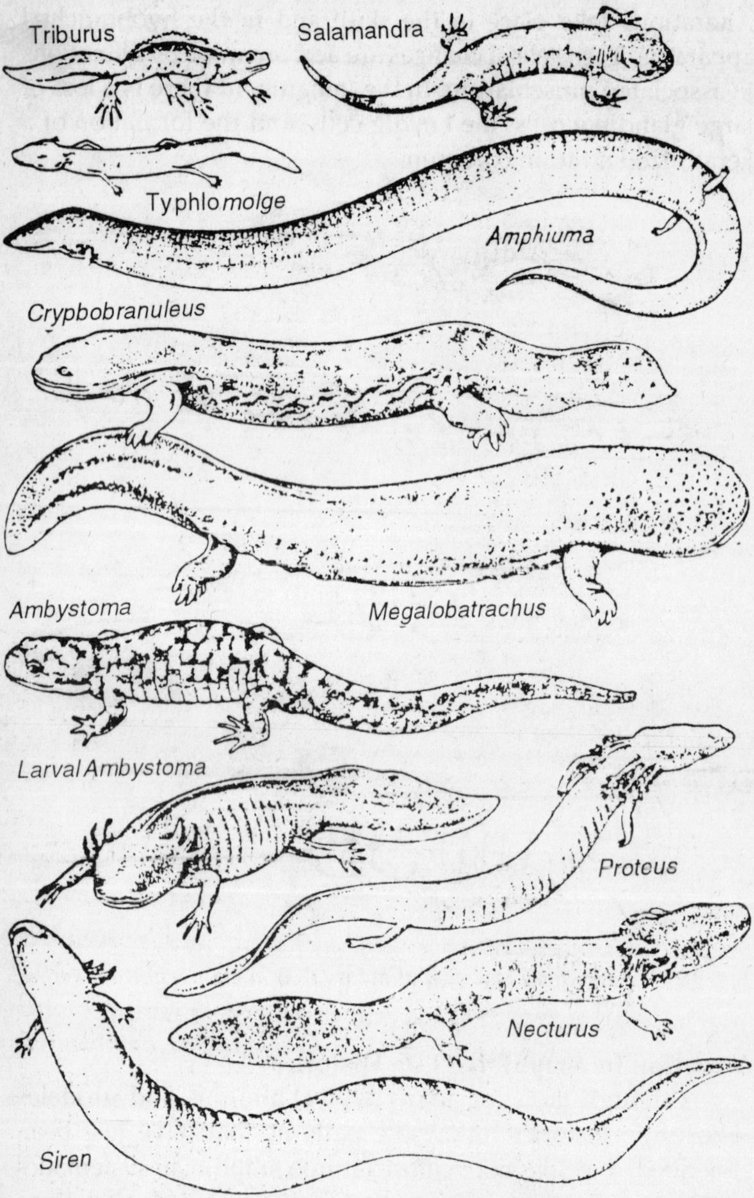

Fig. 9.3. Some of the various urodele amphibians:

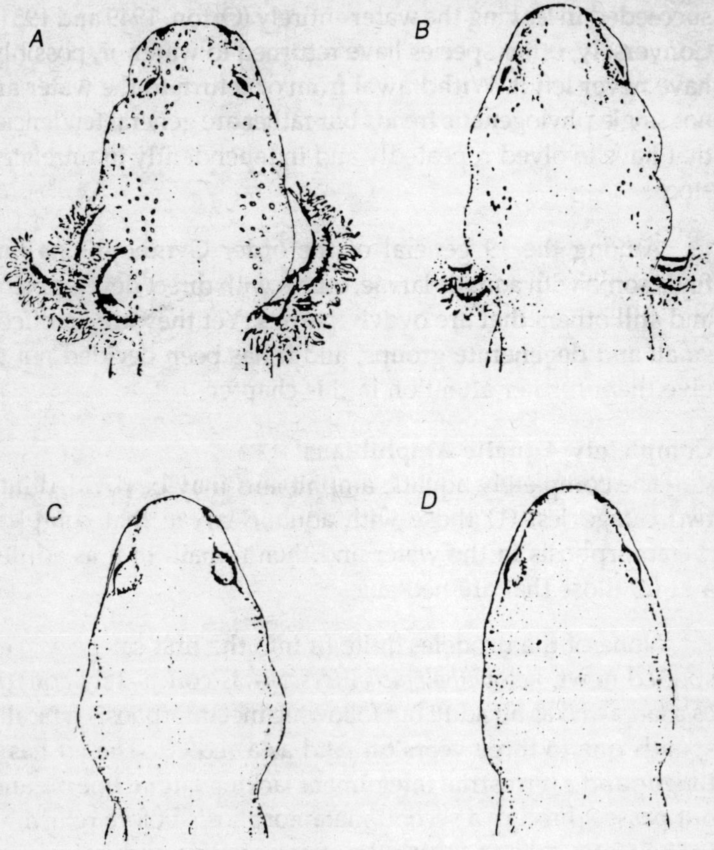

Fig. 9.1. The head region of *Gyrinophilus palleucus* at successive stages of spontaneous metamorphosis.

One tends to think of the amphibians being in the process of withdrawing from the water for life on land. A survey of the various taxonomic groups discloses that there is a fairly complete gradation in a broad range between amphibians that live completely in the water and those that live exclusively on land. The various levels of this gradation, however, neither represent a single orderly sequence nor links in a single evolutionary chain. Instead, it is apparent that on several occasions not only frogs and salamanders of diverse lineage but also the legless members of Gymnophiona as well have succeeded in leaving the water entirely (Orton, 1949 and 1951). Conversely, other species have returned to water or, possibly, have never left it. Withdrawal from or return to the water are not single phylogenetic trends but rather are general tendencies that have evolved repeatedly and independently in unrelated stocks.

Among the 19 general of the order Gymnophiona one finds some with aquatic larvae, others with direct development, and still others that are ovoviviparous. Yet they are, in effect, small and degenerate groups, and it has been decided not to give them further attention in this chapter.

Completely Aquatic Amphibians

The completely aquatic amphibians may be divided into two categories: (1) those with aquatic larvae that complete metamorphosis in the water and then remain in it as adults, and (2) those that are neotenic.

None of the urodeles quite fit into the first category. The spotted newt, *notophthalmus viridescens,* is completely aquatic as a larva and as an adult but following metamorphosis typically spends one to three years on land as a red eft. The eft has a tongue and a terrestrial integument lacking lateral line organs, but passes through a second metamorphosis before returning to the water during which the tongue is lost and lateral line organs are reacquired. *Leurognathus maramorata* of the lungless Plethodontidae is rarely found outside the water but has no special aquatic adaptations as an adult.

Among the Anura the Pipidae, such as the South African clawed toad *Xenopus laevis,* never leave the water. Both larvae and adults are tongueless and have lateral line organs. Occasional aquatic genera are also found among other families such as *Batrachophrynus* and *Pseudobufo* of the Bufonidae and *Telmatobius* of the Leptodactylidae.

Neoteny

There is some confusion in the use of the terms "neoteny"

and "paedogenesis." Neoteny is a condition in which larval characters are retained for prolonged periods of time. Paedogenesis is the process by which larval individuals reproduce. A paedogenetic individual is of necessity neotenic. If a race is to be permanently neotenic it must obviously be also paedogenic.

Paedogenesis is not found among the Anura, although naturally occuring, non-metamorphosing anuran larvae have been reported (Jurand, 1955; Saxen, 1957; Boschwitz, 1957). In some instances the larval periods of certain species or individuals have been prolonged sufficiently for them to be classed as neotenic.

Table 9.1 presents a listing of caudate species reported to be neotenous. All are probably paedogenic as well.

Of the genera listed as permanent larvae, Noble (1931) — upon consideration of such metamorphic features as the development of limbs, maxillary bones, loss of gills, and reduction of branchial arches—proposed that *Siren* (See Fig. 9.3) and *Pseudobranchus* be considered as forms that cease to differentiate at a very early stage of larval life; *Proteus* and *Necturus* as forms that reach a later stage of urodele ontogeny; *Cryptobranchus* as a form that had begun its metamorphosis; and *Megalobatrachus* and *Amphiuma* as forms that have nearly completed their metamorphosis.

The outflow of water from the mouths of urodele larvae through the internal nares (choanae) is usually prevented by the action of a simple flap of connective tissue that serves as a valve. As metamorphosis a more elaborate valve activated by smooth muscle is developed. *Necturus* and *Proteus* have choanal valves and skin of the larval type (with Leydig cell and without stratum corneum), two pairs of gill slits, and three pairs of external gills (no urodele has internal gills). Moving in the adult direction, *Cryptobranchus* and *Amphiuma* each have one pair of gills slits. *Megalobatrachus* has none. These animals

TABLE 9.1

Degrees of Neoteny Among the Urodela

	Family	Genus	Species	Region\	
			Lacertina	Southeastern U.S.	
	Sirenidae	*Siren*	*intermedia*	Southeast and Central U.S.	
		Pseudobranichus	*straitus*	Southeast U.S.	
			maculosus	Eastern U.S.	
		Necturus	*punctatus*	Southeastern U.S.	
	Proteidae	*Proteus*	*anguinus*	Central Europe	
1. Permanent larvae		*Cryptobranchus*	*alleganiensis*	Eastern U.S.	
	Cryptobranchidae		*japonicus*	Japan	
		Megalobatrachus	*davidianus*	China	
			means	Southeastern U.S.	
	Amphlumldae	*Amphiuna*	*tridactylion*	Southeastern U.S.	
		Typhlomolge	*rathbuni*	Texas	
		Haideotriton	*wallacei*	Georgia and Florida	
2. Consistently netenous in natural habitat, but metamorphosis can be induced			*Tynerensis*	Oklahoma	
	Plethodontidae		*neotenes*	Texas	
		Eurycea	*nana*	Texas	
			troglodytes	Texas	
			Gyrinophilus	*palleucus*	Tennessee
	Ambstomatidae	*Siredon*	*pisciformis*	Mexico	

3. Occasionally neotenous in Nature		*Dicampton*	*ensatus*	West Coast of U.S.
			Tigrinuon	United States
	Ambystomatidae	*Ambystoma*	*Talpoideum*	Southeastern U.S.
			gracile	West Coast of U.S.
		Rhyacosiredon	4 species	Mexico
		Notophthalmus	*perstriantus*	Eastern U.S.
			viridescens	Eastern U.S.
			alpestris	Central Europe
	Salamandridae		*cristatus*	Western Europe
		Triturus	*helveticus*	Western Europe
			Taenialus	Central Europe
	Plethodontidae	*Euirycea*	*multiplicata*	Oklahoma, Arkansas, and Missouri
	Hynobidae	*Hynobius*	*lichenatus*	Japan

have not developed eyelids. The larval character of *Siredon* and *Pseudobranchus* is indicated by the absence of hind limbs and the presence of choanal valves, external gills (three pairs in *Siren* but only one in *Pseudobranchus*), and open gill slits. The skin of *Siren* has adult structure, but that of *Pseudobranchus* is larval.

The Thyoid Gland and Neoteny

With the coming of knowledge regarding the relation of the thyroid gland to metamorphosis, questions were raised regarding neotenic forms. Do they fail to metamorphose because of deficient thyroid function or haye the tissues hat ordinarily respond to the thyroid lost their sensitivity to it? In general, the second hormone explanation seems to be most nearly correct, *Necturus* failed more to respond to treatment with thyroxin alone, although it was in combination with adrenalin. The treatment was lethal, and no other metamorphic changes were noted. Etkin (1955) ascribes these results to generalized effects of unfavourable conditions. Gill reduction without other changes was produced in *Siren* and *Pseudobranchus* by treatment with iodothyrine. This, also, may have been a generalized rather than a specific response. Immersion in solutions of desiccated thyroid gland hastened the shedding of skin but had no further effect on *Cryptobranchus*. Equivalent results were obtained in *Amphiuma* and thyroxin. Although *Proteus* gives no metamorphic response to treatment with thyroxin, its skin undergoes changes in the adult direction upon transplantation to metamorphosing salamanders — (*Triton, Salamandra*, or *Ambystoma*).

The placement of the caverniculous salamanders *Typhlomolge* and *Haideotriton* in category 2 is perhaps questionable. In 1957 Dundee immersed a single small (33 mm snout-vent length) specimen of *Typhlomolge* in a 1:1,000,000 aqueous solution of racemic thyroxin. When no results were noted the concentration was increased first to 1:500,000 and then to 1:100,000. Immediately the labial folds, gills, and fins began to undergo gradual atrophy. Maxillae were developed and the beginnings of maxillary dentition were seen. Before the animal died at the end of 19 days, the fins and labial folds were resorbed, and the gills were reduced to mere stubs. The skin was unchanged, and no further metamorphic events were noted. It is possible that if the experiment were repeated using larger animals metamorphosis might be brought to completion. It is well-known that distortion of the metamorphic pattern results when very young larvae are treated with thyroxin.

The evidence favouring the inclusion of *Haideotritión* in category 2 is much stronger. Dundee (1961) immersed five specimens of *Haideotriton* in concentrations of thyroxin ranging between 1:100,000 and 1:2,000,000. All concentrations produced metamorphic changes with about equal effectiveness. In all instances, there was a loss of fins, labial folds, lateral line organs, gills, and the coronoid bone. There were no further changes in the skull, no changes in the skin, and the eyelids were not developed. All five animals died within 25 days of the beginning of the treatment. Like the specimen of *Typhlomolge*, they were relatively small individuals, the larges being 27 mm in length. A mature specimen, the type specimen of *Haideotriton*, was reported by Carr (1939) to be 75.6 mm in length. Pylke and Warren (1958) found that an animal grown to be 43 mm long was still sexually immature. As with *Typhlomolge*, it seems likely that if larger specimens of *Haideotriton* were treated with thyroxin they might pass through a complete metamorphic sequence. It must be conceded, however, that the tissues of both *Typhlomolge* and *Haideotriton* are very resistant to the metamorphic action of the thyroid hormone and that they might be properly classed as permanent larvae.

When immersed in a solution of thyroxin at a concentration of 1:500,000, the Tennessee cave salamander *Gyrinophilus palleucus* was seen after about two weeks to begin metamorphosis with a regression of soft tissues at either side of the tip of the snout. The gills shrank too become mere stubs at the end of about three weeks. By this time the labial folds and tail fin were resorbed, and parasphenoid teeth, maxillary bones and

TABLE 9.2

Examples Illustrating the Trend Toward Direct Development within the Anura

Frog	*Eggs Deposited in*	*Larval Period*	*Metamorphic Modifications*
Rana pipens	Ponds	Typical aquatic	None
Hyla septentrionalis	Water containing cavities of bromeliads	Typical aquatic (rapid development)	None
Hyla faber	Water-filled nests walled off from larger ponds or poles	Typical aquatic	None
Hyperolius tuberillinquis	Vegetation overhanging water	Upon hatching larvae fall into water for typical aquatic larval period	None
Leptodactylus albilabris	Froth-filled nest near pond or pool	Fluctuation of water level washes larvae into pond	None
Hemisus marmoratum	Underground nests near ponds or pools	Female digs tunnel to pond through which larvae wriggle	Lacks external gills but has horny mouth parts
Thoropa petropolitana	Spray-derenched rocks	Tadpoles remain on wet rocks	Mouth adapted for clinging to rocks, tail fins reduced
Leioplema hochstetteri	Damp situations usually (but nort always) floodedd before metamorphosis	Aquatic if flooded	Lacksgills; tail considered to have respiratory function.

Leptodactylus nanus	Froth within earth-covered nest distant from water	Tadpoles remain within cavity	None
Arthroleptella lightfooti	Damp moss	Larvae wriggle on moss for short time before forelimbs emerge	No gills, gill slits, or larval mouth parts; forelimbs covered by operculum.
Cornufer hazelae	Axils for fern fronds	None	No gills, gill slits, or larval mouth parts; forelimbs continuous by perculum that is continuous with abdominal respiratory folds.
Eleutherodactylus nubicola	Damp cavities under stones	None	No gills or gill slits; vestige of larval lower lip and of operculum; tail modified as respiratory structure

teeth were formed. The formation of eyelids took several more weeks, and the differentiation of the tongue was a protracted process that took up to five or six months for completion. Rather surprisingly, in view of the length of time required for complete metamorphosis at this rather high concentration of thyroxin, a few specimens underwent spontaneous metamorphosis in ordinary springwater. The factors that stimulated this spontaneous metamorphosis have not been discovered.

Complete metamorphosis was brought about in *Eurycea tynerensis* and in *Eurycea neotines* within 18 days by immersion in a 1:500,000 solution of thyroxin. The other species of *Eurycea* listed in the second category have not been tested, but it seems reasonably safe to assume that because of their close relation to *E. tynerensis* and *E. neotines* they also are quite responsive to the thyroid hormone. The ease with which metamorphosis is induced in the axolotl *Siredon pisciformis* is common knowledge.

The response of the species listed in category 3 to the thyroid hormone has not been tested in all instances, but it is safe to assume that they are all quite responsive. Metamorphosis occurred quite readily in *Notophthalmus viridescens*, e.g., when neotenic specimens were exposed to quite low concentrations (1:2,000,000) of thyroxin. *Ambystoma gracile* completes metamorphosis at 18 days in a 1:500,000 solution of thyroxin.

Thus, it is reasonably clear that the tissues of the animals in categories 2 and 3 of Table 9.1 are competent to respond to thyroid hormone. But what of their own endocrine glands? For many years it was generally believed that the thyroid gland was congenitally absent in *Typhlomolge rathbuni* , but eventually Gorbman (1957) made a careful histological study and found in each of two specimens around 100 loosely arranged thyroid follicles.

The thyroid of *Gyrinophilus palleucus* typically functions at a low level but produces sufficient hormone to induce metamorphosis when stimulated with exogenous thyroid-stimulating hormone. The thyroid of the axolotl induces metamorphosis when stimulated by pituitary grafts from its typically metamorphosing relative *Ambystoma tigrinum*, but *A. tigrinum* larvae retain their larval form when their pituitary glands are replaced with axolotl pituitaries. In general, these observations and others made on the thyroid glands of facultative neotines indicate that their metamorphic failure results from hypofunction of the pituitary gland.

Withdrawal from the Water

From the standpoint of evolution, several advantages have accrued to those amphibians that have taken up completely terrestrial existence. The survival value of the change is readily recognized when one becomes aware of the wide geographic distribution and the numbers of highly successful species that have acquired this mode of life. The aquatic larva is beset by greater numbers of predators than the developing terrestrial amphibian. On land, parental protection is feasible and often given. The hazards resulting from the drying up of temporary pools and the flooding of mountain streams are eliminated. Metamorphosis is apparently a traumatic process, as evidenced by the fact that in progenies of typical amphibians reared in protected laboratory situations the highest mortality rates are always observed during the metamorphic period. The amphibians that have achieved direct development avoid the rigors that attend the extensive remodelings of typical amphibian metamorphosis.

Adaptations that have resulted in various degrees of withdrawal from the water are found in both the Anura and the Urodela. The range of adaptation is broadest and most complete within the Anura, and we shall give first consideration to them. Gradations of withdrawal can be arranged in two series that are unsequential in the evolutionary sense. One, illustrated in Table 9.2, reaches its culmination with completely direct development in which strictly larval characters are almost entirely eliminated. The other, illustrated in Table 9.3, reaches its peak when ovoviviparity is achieved. The species listed in the left-hand columns of both Tables 9.1 and 9.3 are meant to be representatives of groups of animals that have identical or very similar life histories.

Direct Development Among Anurans

Rana pipiens is representative of all those anurans that have a typical life history in which jelly-encased eggs are laid in some body of water and in which typical aquatic larvae hatch from those eggs to eventually undergo a typical metamorphic sequence. The little West Indian tree frog *Hyla septentrionalis* is representative of those forms that lay their eggs in very small volumes of water such as those found in sections of bamboo or in the cavities created by the branching of leaves from the bases of bromeliads. Their development is morphologically typical but extremely rapid. Such small deposits of water would be likely to evaporate before slowly developing larvae could metamorphose. The South American frogs *Hyla faber*, *H. pardalis*, and *H. rosenbergi* lay their eggs in small nests or puddles of water that are very close to or walled off from larger ponds or pools. *H. rosembergi* has an additional adaptation. Its gills are modified for clinging to the surface film of the water in which it develops.

A large number of species from several genera and families attach themselves to vegetation that overhangs some body of water. In the South American *Hyperolius tuberillinquis* the eggs are contained within a stiff mass of jelly. Oftentimes, as in *Rhacophorus leucomvstax* from the Philippines, the nest is composed of froth beaten up from an albuminous secretion of the female by the limbs of one or both parents. Sometimes the nest remains relatively soft and sometimes it hardens superficially into a crust that resists dehydration. The female of *Chiromantis xerampolina* remains with and moistens her nest until the larvae within it are hatched, otherwise the crust becomes too hard for the larvae to break when the time comes for them to escape. Other forms such as *H. pusillus* attach their eggs individually to leaves or wrap leaves around the eggs, as do the banana frogs of the genus *Hoplophyryne*. In all of these

TABLE 9.3

Examples Illustrating the Trend Toward viviparity within the Anura

Frog	*Embrtis Transported*	*Late Developmental Morphology*
Alytes obstetricans	In strings on hind legs of male until hatching	Typical aquatic larva
Dendrobates auratus	After hatching on land, to water on back of male	Typical aquatic larva
Cryptobatrachus evansi	On back of female until hatching	Typical aquatic larva
Hemiphractus divaricatus	On back of female until development completed	Leaflike gills, simplified mouth parts, rudimentary tail (other species of each
Gastrotheca oviferum	In pouch on back of female until development complete	enus have less yolk, leave the female before metamorphosis, and have better developed larval characters)
Pipa pipa	Individual pockets within thickened dorsal skin of female morphosis completed	Paired spiracles; beaks and labial teeth absent
Hylambates brevirostris	In mouth cavity of female until meta-phosis completed	Typical aquatic larvae except that horny structures are soft and skin is thin
Nectophrynoides tornieri	In uterus until metamorphosis is completed	Gill slits, suckers, horny teety, and beaks are absent, Long, ratlike fail has low lins and a respiratory function

species the larva upon hatching falls into the water and takes up the existence of a typical tadpole.

The majority of the 50-odd species of the genus *Leptodactylus* lay their eggs in froth-filled nests near small bodies of water that fluctuate in depth with rainfall. The nest sites are from time to time flooded and the larvae are routinely washed into the body of water before metamorphosis takes place. *Hylambates natalensis* and a few species from other families are known to build quite similar nests.

With her snout, the female of *Hemisus marmoratum* digs an underground nest chamber and remains with her eggs until they hatch. She then tunnels to the water and is followed by the larvae which, even though they become aquatic, do not develop external gills.

Thoropa petropolitana is representative of a few species of South American frog such as *Cyclorampus pinderi* whose tadpoles are only semiaquatic in that they spend their larval period upon moist rocks and not actually within the water. Their mouths are especially adapted for clinging to the rock, and their tails are modified somewhat as respiratory organs, although gills do develop. *Leiopelma archeyi* from New Zealand is occasionally semiaquatic throughout the larval period. The eggs are laid in a damp situation, but not in the water, and the young swim only if the next happens to be flooded before metamorphosis is complete. It has four pairs of open gill slits, but no gills form. As pointed out earlier, the anuran operculum is derived principally from a transverse dermal fold that grows backward from the hyoid region, but also from a low, ridgelike fold across the anterior part of the belly just behind the forelimb buds. In aquatic larvae these two components are fused together prior to the formation of the forelimb bud. In *Leiopelma* fusion does not take place although the forelimb buds are covered by the anterior components. The tongue forms before hatching occurs. In South American *Leptodactylus* represents a group of several species such as *Zachaenus parvulus* whose larvae

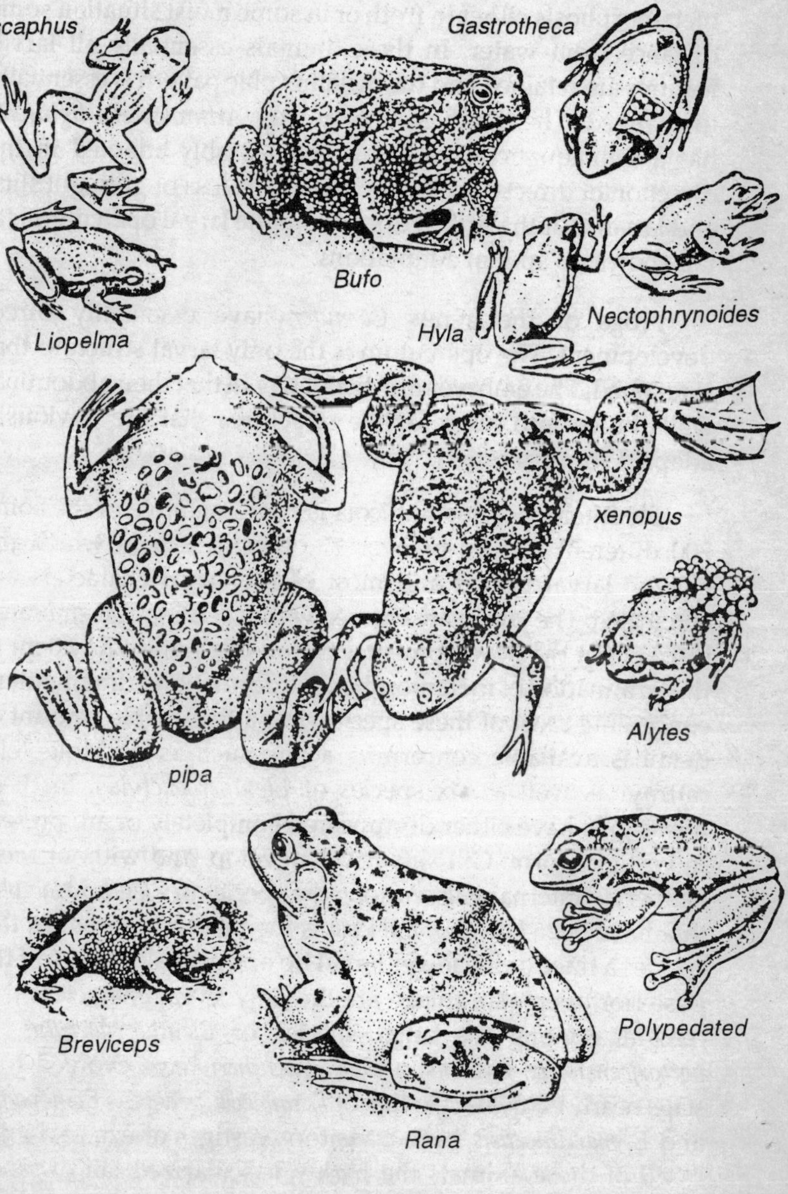

Fig. 9.5. Some of the anuran amphibians.

larvae are semi-aquatic, being laid and left throughout metamorphosis either in froth or in some moist situation some distance from water. In these animals essentially all larval features are retained and the metamorphic pattern is essentially unchanged. The South American frog *Arthroleptella lightfooti* has a semiaquatic larva but is considerably adapted in the direction of direct development with the loss of gills, gill slits, and larval mouth parts. It does retain the larval operculum. Its tail shows no special adaptations.

Frogs of the genus *Cornufer* have essentially direct development. The operculum is the only larval structure that is retained. The embryos are distinctive in that their abdominal walls bulge into huge saclike structures that are obviously adapted for respiration.

Eleutherodactylus nubicola is representative of the some 200 different species of over 15 other genera in which the aquatic larval period and most of the larval characters are eliminated. The young are encased within gelatinous membranes throughout the developmental period to a stage equivalent to the termination of metamorphosis. Information is fragmentary concerning most of these species, but a reasonable amount of detail is available concerning a few such as Arthroleptella rattrayi as well as six species of *Eleutherodactylus*. In these forms gills have either disappeared completely or are present in vestigial form. Gill slits never develop, and without them neither do internal gills. Among the species of *Eleutherodactylus* that have been studied, *nubicola* is the most advanced. In this species a trace of the lower larval lip and a small vestige of the posterior opercular anlage are the only larval characters that remain. Among the other species of *Eleutherodactylus*, *E. portoricensis*, *E. nasutus*, and *E. guentheri* have evolved to a stage nearly equivalent to that of *E. nubicola*, whereas *E. inoptatus* and *E. martinicensis* have transitory vestiges of external gills. In all of these animals the highly vascularized tail expands into a balloonlike structure and serves as an organ of respiration.

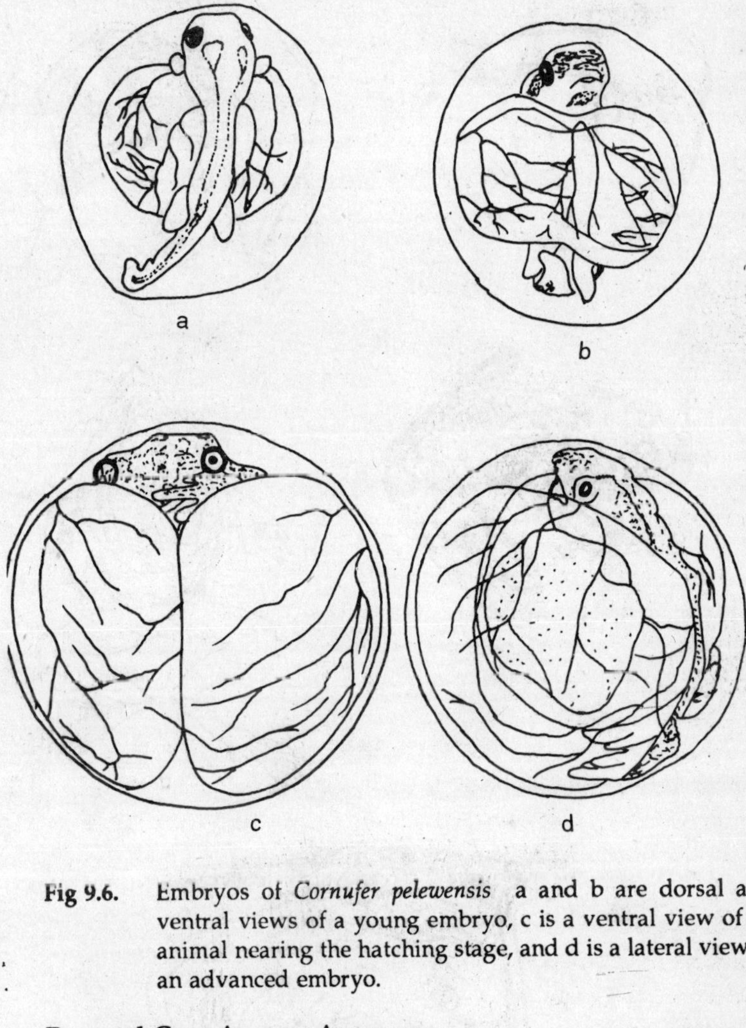

Fig 9.6. Embryos of *Cornufer pelewensis* a and b are dorsal and ventral views of a young embryo, c is a ventral view of an animal nearing the hatching stage, and d is a lateral view of an advanced embryo.

Parental Care Among Anurans

Although they lack fetal membranes, anurans with direct development have otherwise reached a condition of oviparity equivalent to that typical of the birds and of many reptiles. The animals listed in Table 9.3 illustrate another approach to withdrawal from the water. They show a progressive increase in parental care and handling that culminates in true

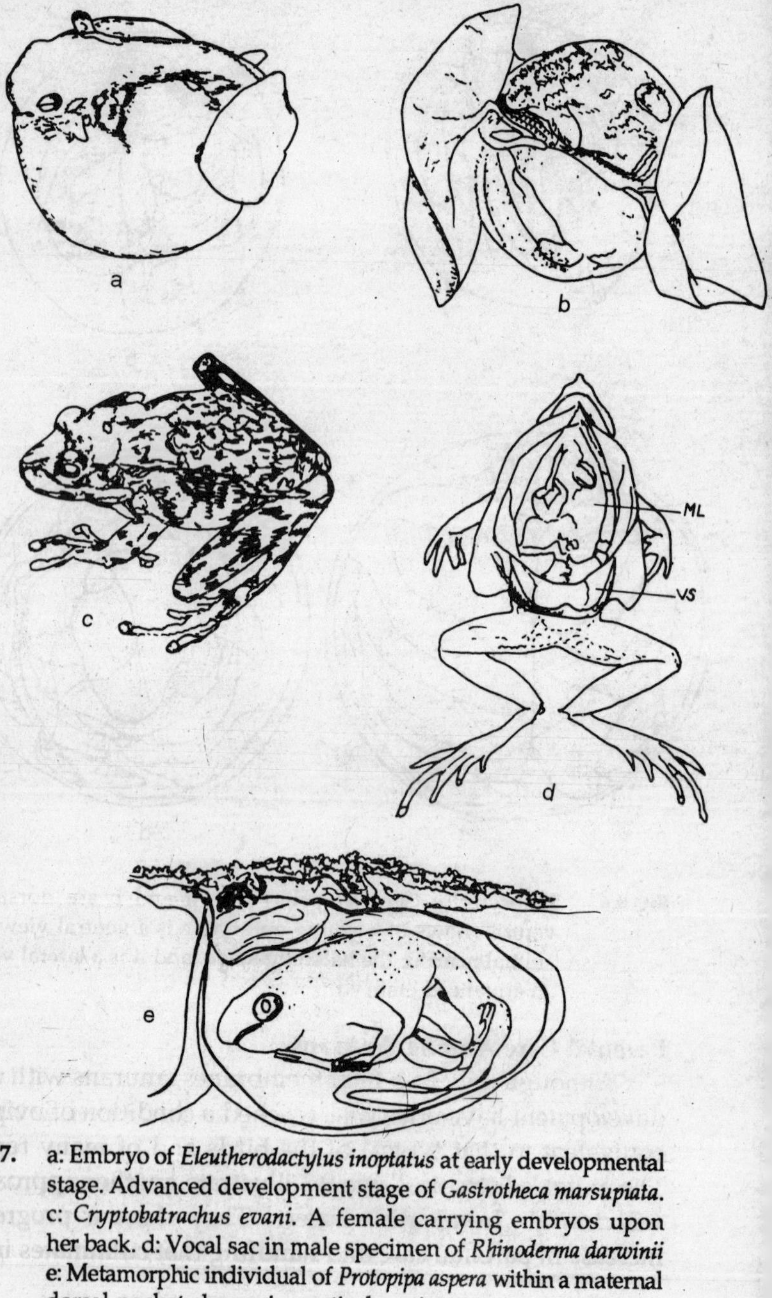

Fig. 9.7. a: Embryo of *Eleutherodactylus inoptatus* at early developmental stage. Advanced development stage of *Gastrotheca marsupiata*. c: *Cryptobatrachus evani*. A female carrying embryos upon her back. d: Vocal sac in male specimen of *Rhinoderma darwinii* e: Metamorphic individual of *Protopipa aspera* within a maternal dorsal pocket shown in vertical section.

ovoviviparity. The "brooding" of eggs is not uncommon among those forms with direct or near direct development. A parent guards the eggs in several species of *Eleutherodactylus*, and the male of *Leiopelma* guards the eggs. It has been suggested by Jameson (1950) that the male of *Eleutherodactylus latrans* wets down the eggs that it guards with urine to prevent their drying out. Some wetting down of the encrusted froth nest of *Chicomantis* by a guarding parent has also been reported by Cochran (1961). The attention of the female of *Hemisus marmoratum* to her young was noted in the preceding section. The animals cited in Table 9.3, however, actually transport their young and have some quite intimate contact with them. The male of *Alytes*, the midwife toad, loops strings of eggs as they are laid around his hind legs and then carries them about, occasionally entering the water so that they are moistened, until finally just before they hatch he enters the water and remains there until the larvae emerge from their membranes.

The males in some species of *Dendrobates* and also those of *Phyllobates* pick up eggs after they have been laid on land and transport them on their backs until some time after hatching takes place. The young of both these genera as well as those of *Cryptobatrachus*, in which the female provides the transportation, are deposited in the water, however, sometime prior to the metamorphosis, *Hemiphractus* and *Gastrotheca*, the "obstetrical frog," which is equipped with a pouch on its back, vary from species to species with regard to the amount of development that takes place while the young are being transported. In the species cited, *H. divaricatus* and *G. oviferum*, transportation continues until metamorphosis is essentially complete but in some other species the larvae swim away before metamorphosis begins. There is also variation in the yolk content of the eggs from species to species, and those larvae with the greater amounts of yolk remain longer on the back of the female.

The Surinam toad *Pipa pipa* is well-known for its adaptation by means of which embryos are transported in individual chambers on the back of the female. The chambers develop

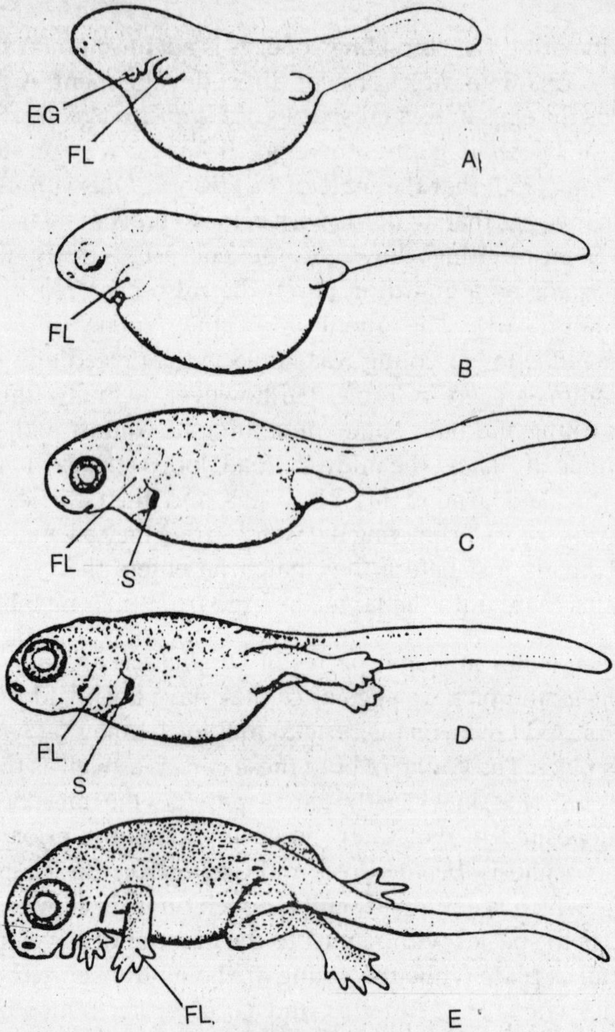

Fig. 9.8. Larvae removed from the oviduct of the ovoviviparous toad, *Nectophyrnoides tornieri*, at various departmental stages. In B the external gills and forelimb buds shown in A have been covered over by the developing operculum. In C operculum formation is complete and the spiracle is evident. In D metamorphosis is near as the forelimb becomes visible through the thin skin of the operculum. In E metamorphosis is underway. The forelimbs have emerged, and the tail fin has been resorbed.

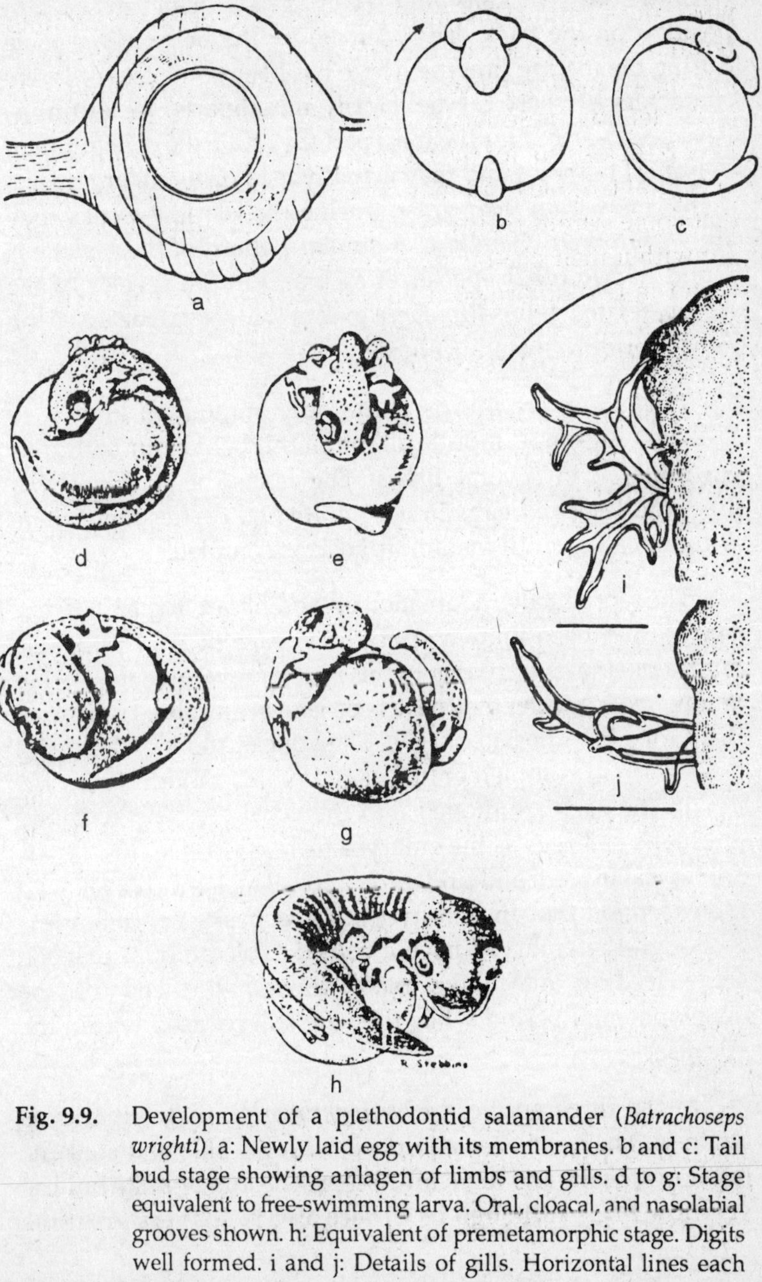

Fig. 9.9. Development of a plethodontid salamander (*Batrachoseps wrighti*). a: Newly laid egg with its membranes. b and c: Tail bud stage showing anlagen of limbs and gills. d to g: Stage equivalent to free-swimming larva. Oral, cloacal, and nasolabial grooves shown. h: Equivalent of premetamorphic stage. Digits well formed. i and j: Details of gills. Horizontal lines each represent 1 mm.

when the skin thickens after the eggs have been pressed into position on the back of the female by the belly of the male during the mating process. There has been some loss of larval characters by these larvae. Horny mouthparts, for example, have been lost, but for the most part they have the configuration of typical tadpoles and may actually feed on small organisms that come within their grasp during the last few weeks they are within their chambers. A similar pattern of life history is found in two other species of *Pipa* and in the species of the closely related genus *Protopipa*, except that the larvae leave the female prior to metamorphosis

Hylambates brevirostris is the only amphibian known to brood its eggs in its mouth, although this behaviour pattern is encountered in several fishes. The female if *H. brevirostris* takes the fertilized eggs in her mouth and carries them there without eating until metamorphosis is complete.

The reproductive adaptation of the Chilean frog *Rhinoderma darwinii* has been known since the voyage of H.M.S., Beagle but is nonetheless striking. In this species several males gather around the eggs after they have been laid and fertilized, guarding them for a brief period of time. Then one or more of the males takes a small complement of eggs into his mouth. From the mouth the eggs are moved into the vocal sac. The vocal sac gradually enlarges as the embryos increase in size. The larvae are typical in form and undergo a typical metamorphosis with the exception that the horny teeth and beaks never harden. Although the embryos have a considerable amount of yolk, some feeding may take place toward the end of the developmental period while they are still encased within the vocal sac.

In the three species of *Nectophrynoides*, ovoviviparity is attained. The larvae are equipped with gill slits and suckers, but horny beaks and teeth are absent. The long, ratlike tail has very low fines. Although its surface area is not very great the tail is well vascularized and obviously has a respiratory function. This is quite apparent in *N. vivipara*. Here the uterus may contain as many as 100 larvae whose bodies tend to be located toward the central part of the uterus and whose elongate tails are arranged so as to have a maximal contact with the uterine wall. Angel and Lamotte (1944) report *N. occidentalis* to be without a spiracle. Orton (1949) states that a spiracle is present in *N. tornieri* but that it is quite small.

Withdrawal from the Water by Urodeles

As in shown in Table 9.4, the movement from the water in the Urodela is roughly parallel to that in the Aruna. Several species (including, oddly enough, the neotenic and otherwise aquatic *Amphiuma*) lay their eggs on land but near water that fluctuates in level so that the embryos are swept into it prior to hatching.

Among the urodeles, direct development is accomplished only in the family Plethodontidae. The large, yolk-laden eggs are surrounded individually with heavy gelatinous membranes. In general, there is no loss of larval characters and no deviation from the typical metamorphic pattern except that the gills are larger and longer than those of aquatic larvae.

In view of its neoteny it is somewhat surprising to find that *Proteus* is often ovoviparous in the lower temperatures of its range. The European newt *Salamandra salamandra* likewise retains embryos in utero until after they hatch, sometimes apparently stimulated by low temperatures, throughout metamorphosis. *Salamandra atra* is consistently ovoviviparious, as are the plethodontids, *Hydromantes* and *Oedipus*. As in the ovoviviparous urodeles, with direct development there is no loss of larval characters in ovoviviparious urodeles. The intrauterine larvae of *Salamandra*, for example, develop rudimentary balancers and lateral line organs (Noble, 1931). The elongate gills of these creatures function in the exchange of respiratory gases between maternal and larval bloodstreams.

TABLE 9.4

Urodeles with Terrestrial Paterns of Life History

	Family	*Genus*	*Species*	*Region*
Eggs laid on	Ambystomidae	*Ambystoma*	*opacum*	Eastern U.S.
land but	Amphiumidae	*Amphtuma*	*Tridactylium*	Southeastern U.S.
larvae		*Desmorgnathus*	9 species	Eastern U.S.
aquatic	Plethodontidae	*Hemidactylium*	*scutatum*	Notheast U.S. and Southeast Canada
		Stereochilus	*marginatum*	Southeastern U.S.
Eggs laid on		*Batrachoseps*	3 species	West Coast of U.S.
land with	Plethodontidae	*Aneides*	16 species	Appalachian Mountains and West Coast
direct		*Plethodon*	16 species	U.S. and Canada
Development		*Ensatina*	*escholtzi*	West Coast North America
	Proteidae	*Proteus*	*anguinus* (at lower temperatures)	Central Europe
Eggs retained *in utero* during development	Salamandridae	*Salamandra*	*Salamandra*	Central Europe
			atra	Central Europe
		Hydromantes	5 species	Southern Europe and California
	Plethodonotidae	*Ocdipus*	30 species	Central and South America

The Role of the Thyroid Gland in Direct Development

In a preceding section the involvement of the thyroid gland in amphibian neoteny was commented upon. Its function in direct development is also a matter of considerable interest. As early as 1917 Hoskins and Hoskins showed that the course of typical metamorphosis was halted by removal of the thyroid gland. In 1925, Allen found some difference in *Bufo* and *Rana* in the degree to which limbs will grow in the absence of thyroid stimulation. In *Bufo* the hind limbs of thyroidectomized larvae may reach a length of 8 mm, whereas those of *Rana* remain as small buds. It is well-known that administration of thyroid hormone to normal tadpoles brings about precocious metamorphosis. It seems possible, then, that direct development may consist of a telescoping of larval stages resulting from a precocious metamorphosis brought about by an early and intensive functioning of the thyroid gland, by an usually high sensitivity of the tissues to a relatively low thyroidal activity, or by the emancipation of the tissues from the need for stimulation by the thyroid hormone. Histological observations made by Lynn in 1936 indicated that in *Eleutherodactylus* the thyroid began to function early in the developmental period and therefore might be responsible for a precocious metamorphosis. Similar conclusions were reached by Brink in 1939 with regard to *Arthroleptella bicolor*, but Dent found that the thyroid differentiates late in the intraoval life of the terrestrial salamanders *Plethodon cinereus* (1942) and *Aneides anaeus* (1954).

Embryos of neither *Plethodon* nor *Eleutherodactylus* have survived attempts to remove the anlage of the thyroid by surgical procedures. In 1947, however, Lynn treated embryos of *Plethodon cinereus* with the goitrogen thiourea, blocking thyroid function by chemical means. Those embryos completed their development except for the resorption of the long and branched gills. Similarly, Lynn (1948) and (Lynn) and Peadon in 1955 reported results of treating with goitrogens *Eleutherodactylus ricordii* and *Eleutherodactylus martinicensis*. Three metamorphic features were affected by the treatment. The pronephoroes did not degenerate, the tail was not resorbed, and the egg tooth the animals develops just prior to hatching did not drop off as it ordinarily does. Administration of thyroxin to *Eleutherodactylus martinicensis* caused precocious degeneration of the pronephroes and resorption of the tail. Thus, it seems clear that the final steps of direct development are under control of the thyroid but that in the course of evolution the greater part of the ontogenetic sequence has been freed from thyroid control in *Plethodon* and in *Eleutherodactylus*.

Environmental Factors in Metamorphosis

Dodd and Callan (1955) found sexually mature neotenic individuals within an otherwise typical population of *Triturus helveticus* congregated in a pond of the count of Fife in Scotland. As indicated in Table 9.1, neoteny has been reported to occur occasionally in several species of *Triturus*. These specimens of *T. helveticus*, however, were unique in that their thyroid glands were goiterous and tremendously enlarged. This condition could hardly be attributed to a lack of iodine in the environment since the pond is situated within a few miles of the North Sea, and the iodine content of drinking water in the nearly town of Crail in 5 μg/1. Dodd and Callan noted that kale and turnips were cultivated in fields in the vicinity of the pond. The goitrogenic effects of the brassicas is well-known. Rabbits were frequently seen in the vicinity of the pond, and rabbit feces accumulated on the slopes from which water drained into the pond. Dodd and Callan concluded that this drainage water should contain the "brassica factor" at times when the rabbits were feeding on turnips and kale and that larvae overwintering in the pond might well, as a result, become goiterous and neotenic. Although it is interesting, it is unlikely that this condition described by Dodd and Callan for *T. helveticus* obtains generally among other occasionally neoteni forms.

Notophthalmus viridescens is neotenic in the lowlands of Louisiana, on Cape Cod, and on Long Island; but the majority of the neotenous individuals among the species listed in category 3 of Table 9.1 live in lakes located at altitudes of several thousand feet, as does *Siredon pisciformis*, the axolotl cited in category 2. Snyder (1956) has made comparative study of the Pacific Coast salamander *Ambystoma gracile* from ponds found at sea level and from other ponds at high elevations (4,300 to 5,500 ft.) in Mount Ranier National Park. At sea level a high proportion of larvae transform into immature adults at one year of age, and most of the remainder transform at two years of age as sexual maturity is attained. A small number persist as neotenic larvae. Adult animals are found in the montane habitats, but metamorphosis appears to be rare and most of the mature animals are neotenic. Snyder states that among amphibians cold either inhibits the release of thyroid hormone or the ability is tissues to respond to its presence, and Huxley (1929) has demonstrated that below 50°F the tissues of tadpoles are not stimulated by thyroidal material. Frieden has studied this effect extensively. Snyder states further that younger larvae respond more readily than older animals to the metamorphic effects of the thyroid hormone. This contention is supported by observation of Lipchina (1929) made on the axolotl. At high elevations the annual period of activity and growth lasts only about three months, whereas at sea level larvae may be active and feeding throughout the year. Snyder suggests that since the montane larvae reach metamorphic size at a later age they are possibly less responsive to the thyroid hormone or perhaps produce less thyroid hormone and thus do not metamorphose. The presence some neotenic animals at sea level is accounted for by Snyder on the grounds that eggs are laid over a period of five to six weeks and that perhaps the last eggs give rise to animals that overwinter as larvae and reach the age at which they are resistant to the thyroid hormone. Some brief experiments of a preliminary nature appeared to give support to this hypothesis, but further studies should be made.

From an evolutionary point of view its seems reasonable that the selective pressures would be greater on the barren icy borders of a mountain lake than within its waters. Similarly, in the caves inhabited by most of the animals of category 2 in

Table 9.1 it is usually quite apparent that food is more abundant and other conditions are more favourable to the salamander within the water than outside of it. Very likely it is these selective pressures that have brought about the rise of neoteny in caves and in mountain lakes.

Pedogenesis has never been found to occur among the Anura, but the larval period in some frogs varies greatly in length. Temperature appears to be the predominant factor involved. *Rana clamitans*, the green frog, is usually said to metamorphose after spending one winter as a larva, but Ting (1951) showed that tadpoles of *R. clamitans* metamorphosed in 92 days in the laboratory, and Martof (1952) found that great numbers of larvae metamorphosed in August from a pond that had been dry the preceding summer. He thought that overwintering larvae come from eggs laid late in the season. The bullfrog *Rana catesbeiana* also has a larval period of variable length, as sown in Table 9.5, which is taken from a paper by Willis, Moyle and Baskett (1956). These figures indicate that the larval period increases with the length and severity of winters.

TABLE 9.5

Age of Bullfrog Tadoles At Transformation in Various Parts of the United States[a]

Locality	*Number of Winters of Life Before Transformation*				*Reference*
	0	1	2	3	
New York			X	X	Wright (1914)
Iowa			X	X	Carlander et al. (1950)
California			X		Storer (1922)
Iowa		X			Klimstra (1949)
Florida		X			Fla. D. Ag. (1952)
Louisiana	X	?			George (1940)
Gulf States	X	X			Viosca (1934)

The length of the larval period may also be increased in urodeles without the occurrence of paedogenesis. Many individuals of *Eurycea bislineata* metamorphose in their second year, but some (presumably those from eggs laid late in the first year) metamorphose in the third year . In southern Europe *Triturus vulgaris* metamorphoses in the fall of its first summer, but in northern Russia it hibernates for one winter as a larva , no doubt because the summers in the latter region are not long enough or warm enough for it to complete its larval development in one of them.

In looking about for environmental features that may be responsible for neoteny in any given group of amphibians one always given consideration to the iodine content of the water. Actually, no one has yet reported finding neotenous larvae in water with an iodine content too low to support metamorphosis in other amphibian species. Elair (1961) reported that one of two specimens of neotenic *Gyrinophilus palleucus* metamorphosed after three months in a rather strong solution of sodium iodide, but the other was still uncharged after six months, and as was pointed out earlier Dent and Kirby-Smith (1963) found that specimens of this species occasionally metamorphose spontaneously in spring water under laboratory conditions. Although the induction of metamorphosis in the axolotl by the application of iodine has been reported, the amount given was relatively large. It seems safe to conclude that low iodine content of water is not ordinarily a cause of neoteny.

It is to be expected that larvae deprived of food would soon stop growing and differentiating. As far back as 1878 Yung reported that amphibian larvae are highly susceptible to nutritional deficiency and that their development is conditioned by the nature of the food. In 1887, Barfurth noted that under certain conditions the development of tadpoles was accelerated by starvation. This phenomenon was investigated in detail by D'Angelo, Gordon and Charipper (1941). They found that up to a critical period occurring in the early stages of hind-limb development metamorphic progress is slowed down and then halted by inanition so that tadpoles can be held for considerable periods of time in a "stasis" condition. Having passed the critical stage, the rate of metamorphosis indeed is increased by inanition. Possibly inanition accelerates the degenerative phases of metamorphosis. D'Angelo, Gordon, and Charipper observed that the thyroid and pituitary glands of starved animals underwent atrophic and degenerative changes. Stasis tadpoles metamorphose readily after either immersion in thyroxin solutions or injection with pituitary material. The thyroid glands of stasis animals that are metamorphosing after treatment with thyroxin are histologically inactive, but those of pituitary-injected tadpoles give histological evidence of marked activity. On the basis of these observations it was concluded that the failure of starved animals to metamorphose is directly related to a decreased production and release of thyrotropic hormone from the anterior hypophysis.

Any naturally occurring inanition would be likely to be associated with crowding, and crowding itself appears to have a direct effect on the growth of amphibian larvae. It was Yung (1885), again, who first made observations on the relation of crowding to the delay in tadpole metamorphosis. His work was followed by that of others; notably, Adolph in 1931 and more recently Richards (1958 and 1962), Rose (1960) and Akin (1966). Adolph saw that uncrowded animals reaches a much larger size than crowded ones and that they metamorphosed in a much briefer period of time. Rose observed that whenever a group of *R. pipiens* embryos were put in a rather confined space all began to grow after hatching, but at different rates. Those that grew more rapidly at first and became larger than their bowl mates continued to grow if the water was changed daily, whereas those that lagged behind stopped growing and failed to eat even when abundant food was supplied to them. On the other hand, the stunted animals grew if they were removed and put into other containers, giving greater "lebensraum." Richards (1958) showed that if small tadpoles were put into culture water in which large tadpoles had been growing the growth of the small tadpoles was completely or almost completely inhibited, indicating that some sort of inhibitory product is thrown off into the medium by the larger tadpoles. Further, the inhibitory effect could be removed by heating the culture water to 60°, by centrifugation, by sonification, or by filtration. Thus, the inhibitory product is a rather large particle.

In an aquarium where both large and small animals were present and the small ones were growing poorly and dying out, the largest animals were just as large as large animals that had been growing in isolation (Rose, 1960). This indicates that the large animals were unaffected by anything produced by the small tadpoles or that the small ones were neither producing stimulatory nor inhibitory substances. The growth of small tadpoles was supported, on the other hand, by culture water from large tadpoles after it had been treated with a proteinase. At least some of the inhibitory material, then, must be a protein. Richards (1962) reported that the inhibitory agent is associated with a type of algal cell in *R. pipiens*. Akin (1966) showed that the agent is elaborated by the posterior half of the gut in the growing tadpole. The algae pass through the gut and apparently transport the agent to the water. Akin also confirmed earlier reports that the agent is largely species-specific.

There are not reports of the inhibitory effects of crowding in nature, but such effects must exist since in the laboratory one rapidly growing large tadpole can appreciably retard the growth of small tadpoles in as much as 75 liters of water.

There have been reports that various electrolytes have accelerating or inhibitory effect on metamorphosis. Lynn and Wackowski (1951), however, are of the opinion that these apparent electrolytic effects are caused by changes in the hydrogen ion concentration of the medium rather than by any specific effects of the ion in question. Marzulli in 1941 confirmed earlier observations of Rosen (1938) to the effect that acidity below pH 4.8 and alkalinity above pH 11.0, respectively accelerate and retard the metamorphosing action of thyroxin. Marzulli further demonstrated that if thyroxin is injected its effects are not dependent on the pH of the culture medium.

He concluded that the greater effectiveness of thyroxin dissolved in an acid culture medium results from the fact that it is taken up more actively by the tissues at acid rather than alkaline pH levels. For further discussion of the uptake of thyroxin.

Disclos (1959) reported increased metamorphic rates among tadpoles of *Alytes obstetricans* given supplemental illumination, and Guyétant (1964) found that growth and metamorphosis were accelerated in larvae of *R. temporaria* maintained in constant illumination at 2,000 and 4,000 lux, whereas tadpoles kept in permanent darkness grew more slowly. It seems likely that the effects of illumination upon development will be found to be quite complex. Possibly they are mediated through the neuro-secretory pathways of the hypothalamus. This area of investigation is relatively untouched and may be the source of very interesting results in the future.

10

Hormonal Control

If instead of reviewing the published endocrine studies on amphibian metamorphosis we ask questions concerning the endocrine control of the process from the viewpoint of developmental physiology, we become impressed more with the lacunae in our knowledge than with the strength of the separate threads. In any event the general endocrine interrelations in this field have been reviewed in some detail Here we propose to examine metamorphosis from the viewpoint of the physiology of animal development rather than to inquire into the role of the separate endocrine glands. Before we can do that, however, we must summarize the main outlines of the endocrine relations among vertebrates generally in order to provide the framework within which we can ask our questions concerning control of metamorphic change by the hormones.

Endocrine Relationships

No detailed citations to the literature will be given for this discussion of endocrine interrelationships, but such references can be found in recent summaries of the field such as the recent volumes on *Neuroendocrinology* and *The Pituitary Gland* and in standard textbooks of endocrinology.

The thyroid hormones (TH) are iodine-containing derivatives of tyrosine. The principal hormone is thyroxine. The thyroid gland has an iodine-trapping mechanism whose efficiency as determined with radioiodine (I^{131}) is one measure of thyroid activity. The histological picture presented by the thyroid is also a clear indicator of its physiological state.

Abundant cytoplasm and release of colloid from the follicles indicate active glands. However, where iodine uptake is prevented by goitrogens, a histological picture of activity is associated with the failure to release potent hormone.

The thyroid gland can function autonomously at a low level, but higher levels of activity are dependent upon activation by a hormone, thyrotropin, or thyroid-stimulating hormone (TSH), produced by the anterior lobe of the pituitary. The level of activity of the pituitary-thyroid (PT) axis is under control through two known mechanisms.

One of these is by way of feedback of thyroid hormones that inhibit the activity of the TSH cells, thus stabilizing the PT axis at a particular level. There is good evidence that this negative feedback of TH upon TSH production operates primarily at the pituitary level, directly inhibiting the TSH cells.

The second mode of regulating TSH activity is by way of a thyrotropin releasing factor (TRF) produced in the hypothalamus. TRF is presumed to be a neurosecretory material reaching the anterior pituitary by way of the pituitary portal veins that drain the median eminence of the hypothalamus. At present the chemical nature of TRF has not been clarified nor have the neural cells from which it arises been identified. It is believed, however, to differ from the peptides that constitute the classical neurosecretory substances produced in hypothalamic nuclei (preoptic nucleus in amphibians) and stored in the neural lobe of the pituitary. In stimulating TSH production, TRF may be thought to act by partially desensitizing the pituitary to the negative feedback action of TH and thereby permitting the level of the TP axis, the so-called thyrostat, to rise.

A word of clarification is needed with regard to the concept of neurosecretion. This has undergone marked development in recent years Originally neurosecretion was identified by characteristic, although not specific, staining reactions in light microscopy, i.e., neuronal granules positive to aldehyde fuchsin or chrome alum haematoxylin. Physiological, histochemical, and electronmicroscopical evidence has compelled the recognition that such material is related primarily to the pars nervosa although also occurring in the median eminence (Oota and . This latter region, however, has been shown, to be rich in monoamines and in peptides differing from those of the nervoasa. In the present discussion the neurosecretory system will be considered to include in addition to the classical neurosecretory pathway to the posterior lobe of the pituitary:

1. Neurons that produce other specific chemicals released into the bloodstream in the median eminence.
2. The median eminence with its primary capillary bed intimately related to neurosecretory fibers.
3. The portal veins that drain blood from the primary capillary bed of the median eminence into the secondary capillary bed of the pars anterior of the pituitary .

These endocrine and neuroendocrine interrelations are applicable to amphibian metamorphosis. As early as 1912, Gudernatch reported that the feeding of thyroid-gland preparations to young tadpoles precipitated metamorphic changes. It was soon demonstrated - particularly by M. Allen, Hoskins, and Hoskins and P. E. Smith — that the pituitary gland of the tadpole controls thyroid activity and that the latter is the immediate agent of metamorphic change. The neuroendocrine relations governing metamorphosis were worked out more recently. The evidence is convincing here, too, that the hypothalamus influences pituitary TSH activity by way of a TRF passed into the capillaries of the median eminence.

Another pituitary hormone that recently assumed a role in our understanding of metamorphosis is prolactin, which like TSH is a protein product of the anterior lobe of the pituitary. It is known to be active in stimulating body growth in several vertebrates and has recently been found to have antithyroid activity in tadpoles. In mammals and probably in amphibians prolactin appears to be under inhibitory rather than stimulatory control by the hypothalamus.

The Metamosis in a Typical Anuran

However, the endocrine analysis with which we are concerned here has been carried out largely on a few species of anurans. To follow this analysis it will be most useful to have a more exact and quantitative delineation of the developmental pattern relative to metamorphosis in a typical Ranid, *Rana pipiens.* The principal points to be noted here are as follows. The first post-embryonic period characterized by much growth with very little change in form. This is here designated the growth or premetamorphic period and lasts about seven weeks at conventional room temperatures (22 to 25°C). This is followed by the period designated prometamorphosis and characterized by differential growth of the hind legs. During this period of about three weeks, the legs grow from about 2 mm to about 20 mm, whereas the body growth increases only from about 55 mm total length (body length, 18 mm; tail length, 37 mm) to 65 mm body length, 21 mm; tail length, 44 mm). It is easy to follow the progress of the animal through prometamorphosis by nothing the ratio of hind-leg length to body length (HL/BL).

The animal has definitely entered prometamorphosis when the HL/BL ratio exceeds 0.2 and it reaches the end of that period when this ratio approaches 1.0. During the latter part of prometamorphosis certain minor morphological changes may be observed in the intact animal. For example, the resorption of the anal canal piece of ACP (the basal lobe of the ventral fin that carries the anal opening out into the left side of the tail) occurs when the HL/BL ratio is about 0.6 In a day or two thereafter at a ratio of about 0.8 the first signs of the degeneration of the opercular skin over the gill chambers (the skin window for the forelegs, SWFL) can be seen on the right side. The corresponding change on the left is the reduction of the tubular

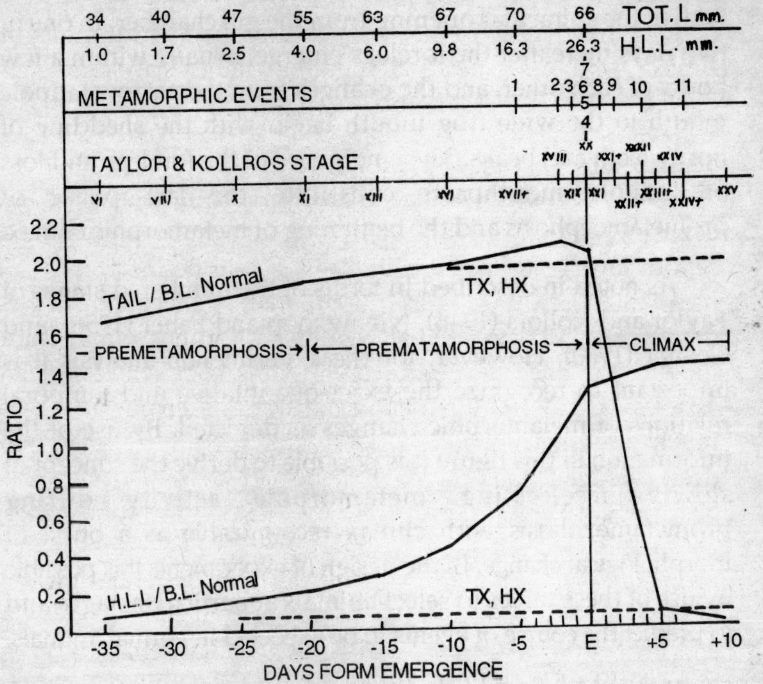

Fig. 10.1. Pattern of metamorphosis in *Rana pipiens*. Data from one batch of normal animals raised at 23°C = 1 shown by solid line. Comparable data for thyroidectomized animals (TX) and hypophysectomized animals (HX) are shown by broken lines. The metamorphic events indicated in the figure by number are as follows: 1. Anal canal piece-first definite reduction. 2. Anal canal piece reduction completed. 3. Skin window for the forelegs clearly apparent. 4. Loss of 2d (both) beaks. In experimental animals this is the most satisfactory criterion for the beginning in climax. 5. Emergence of first foreleg to appear. 6. Emergence of second foreleg (E). 7. Mouth widened to level of nostril. Events 4, 5, 6 and 7 usually occur within a period of 24 hours. 8. Mouth widened to level between nostril and eye. 9. Mouth widened to level of anterior edge of eye. 10. Mouth widened past level of middle of eye. 11. Tympanum definitely recognizable. By use of the data of this table it is possible to classify individual animals in terms of days before or after the beginning of climax (E).

wall of the spiracular opening from the gill chamber. In one or two days thereafter the forelegs emerge, usually within a few hours of each other, and the changes from the narrow tadpole mouth to the wide frog mouth begin with the shedding of horny teeth and beaks. The emergence of the forelegs and loss of tadpole mouthparts constitute the last phase of prometamorphosis and the beginning of metamorphic climax.

As noted in described in terms of the numbered stages of Taylor and Kollors (1946), Nieuwkoop and Faber (1956), and Gosner (1960). However, for the experimental analysis it is important to recognize the exact quantitative and temporal relations of metamorphic changes as depicted. By use of the information in this figure it is possible to derive the concept of slowly accelerating metamorphic activity during prometamorphosis with climax recognizable as a burst of morphological change. In the design of experiments it is possible by use of these tables to select animals at particular stages and to predict the course of events to be expected in normal animals.

It should be noted that the widely used stages of Taylor and Kollros describe the entire range of tadpole development and are not directed at the analysis of metamorphosis per se. They do not recognize the differential acceleration of hind-leg growth as the initiation of metamorphosis but regard this process as beginning in what we designate here as late prometamorphosis. They also do not differentiate a climax phase. Descriptively, the procedure of Taylor and Kollros is highly useful, but it obscures the experimental analysis. As will be seen later, early leg growth (stage XII to XVIII) is dependent upon activation of the myroid gland and is thus considered here to be part of metamorphosis. Furthermore, the endocrine balance undergoes a profound change at the beginning of climax so that the precise delineation of this phase is essential for an adequate understanding of the hormonal control of metamorphosis.

The most important consideration that emerges from this

analysis is that the postembryonic developmental cycle of a typical anuran involves three phases: (1) a growth phase characterized by rapid growth and little morphological change, (2) a prometamorphic phase with reduced body growth rate and with morphological changes proceeding at a progressively accelerating pace, and finally (3) a phase of metamorphic climax in which body growth has ceased and differentiative changes proceed with extreme rapidity. We may center our analytic discussions, therefore, upon the special endocrine characteristics of each phase and the regulation of the transition from one to the next.

The Role of the Endocrines in the Growth Phase of Tadpole Development

It is obvious that at least morphologically all the elements necessary for metamorphic activity are present in the early tadpole, the hypothalamus, the pituitary, the thyroid, and the tissues. We may then ask: Why does not the animal transform immediately? Many possibilities suggest themselves, and we may consider the more plausible of these seriatum:

1. Are the tissues initially insensitive or relatively insensitive to thyroid hormone? Such a change in sensitivity might determine the time of initiation of metamorphic activity. The evidence, however, seems to indicate clearly that this is not the case. It is indeed true that the tissues of the embryo seem to be insensitive to thyroxine as seen in the often repeated observation that embryos develop normally in thyroxine solution and do not show metamorphic change (e.g., leg growth or tail resorption) until reaching the definitive tadpole stage (stage I)—for earlier literature see Etkin, 1955. On the other hand, it is also clear, as in Gudernatch's original experiments, that the postembryonic tadpole's tissue is capable of giving metamorphic responses to thyroid hormone. In a brief study reported only in abstract form, Etkin (1950) attempted to determine more precisely just when the tissues acquire thyroxine sensitivity. He did this by exposing embryos of *R. pipiens* to immersion in

strong thyroxine solutions (or implantation of thyroxine crystals), beginning at successively later stages. He observed that embryos exposed before embryonic stage S-23 did not develop metamorphic changes any earlier than did those whose exposure began at stage S-23. On the other hand, those exposed at later stages showed the first evidence of response at successively later periods. He concluded, therefore, that the tissues first became sensitive at about stage S-23 and are not influenced by exposure at earlier stages. There is a latent period of about three days in the appearance of metamorphic change, even when late tadpoles are exposed to strong thyroxine—1 part per million (ppm). The first changes do not appear in the above experiment until the animals are in stage S-25. Recent experiments by Prahlad and De Lanney (1965) with embryos of the axolotl likewise indicate early acquisition of thyroxine sensitivity. Furthermore, the pattern of change in the embryos of *R. pipiens* exposed at stage S-23 is the same as that in animals exposed, later. This suggests that the general tissues, as observed. externally, acquire sensitivity at the same time and approximately to the final degree.

There have been many attempts to show that sensitivity of tissues to thyroxine increases during tadpole development (e.g., Heady and Kollros, 1964). However, the present author does not regard the evidence available on this point as satisfactory for two reasons: (1) When, as in the experiments that claim to demonstrate changes in sensitivity, normal tadpoles of different ages are compared in their reactivity, the older animals start from a more advanced baseline since their tissues have been exposed for a longer time to conditioning by low levels of thyroid hormone present in the tadpole. (2) The differences in sensitivity reported are small, less than a factor of two times and therefore as compared to the changes in thyroid level can play only a minor role in determining normal metamorphic change. In some cases differences in time of first appearance of metamorphic change in animals treated as embryos have been interpreted as indicating differences in time of acquisition of sensitivity. However, such differences may represent merely differences in latent period of response. In summary, then, except for certain nervous structures, the tadpole's tissues appear to be ready to respond to the metamorphic stimulus throughout the growth phase of development. Changes in sensitivity play a minor, if any, role in determining metamorphic pattern.

2. A second possibility to be considered is that the thyroid gland may not be capable of forming the necessary hormone during the growth phase. However, the earliest studies have shown that the thyroid gland differentiates normally in the absence of the absence of the pituitary. In the normal animal, moreover, the gland forms detectable levels of T_4 in the early tadpole stage (16 mm total length in *R. pipiens*), as shown by Flickinger (1964). The tadpole's thyroid also responds to pituitary TSH at all stage of tadpole development and even in the embryo. Therefore, there is no reason to suspect that the thyroid of the tadpole is incapable of responding appropriately to a metamorphic stimulus during any part of the growth phase. We must look higher than the thyroid in the hypothalamus-pituitary-thyroid axis for the explanation of the failure of the tadpole to metamorphose during this period.

If we ask next whether the pituitary is capable of effective levels of TSH production during this period, we must again answer that it is. The most direct and satisfactory evidence for this would be the demonstration of the responsiveness of the tadpole pituitary to a hypothalamic TRF factor. However, as is discussed below, evidence from this source is not yet available. But other evidence does indicate clearly that the tadpole's pituitary is capable of producing an effective TSH during tadpole life. One such bit of evidence is the demonstration that when the pituitary (adenohypophyseal) primordium an thyroid primordium are placed close together in the embryo, the thyroid is precociously activated (Etkin 1939 and 1966a). This shows that even at early stages the pituitary is capable of producing an effective TSH. The evidence derived from the effects of goitrogens in tadpoles further demonstrates that this capacity for producing TSH continues through tadpole life. The administration of such goitrogens leads to hypertrophy of the thyroid (for literature, see Etkin, 1964, and Hanaoka, 1967). this action, like the similar action in mammals, works by preventing the iodination of the molecule in the gland and, therefore; renders its hormone production ineffective. The lack of thyroid hormone in the goitrogen-treated tadpole removes the feedback inhibition normally exerted by thyroid hormone upon TSH production. The excess TSH produced by the pituitary of the goitrogen-treated tadpole stimulates the thyroid to increased growth and to the abundant secretion of its physiologically ineffective product. We have recently found this goitrogen effect to obtain even in the early tadpole ('unpublished results). Thus, we are led to regard the pituitary of the early tadpole as capable of producing the TSH necessary for the induction of metamorphosis but restrained from doing so because of the inhibition of its TSH-secreting cells by negative feedback from thyroxine produced by the thyroid. The premetamorphic period may be regarded as one in which the pituitary-thyroid axis is maintained in a steady state at an extremely low level of activity because of the great sensitivity of the TSH cells of the pituitary to negative thyroid feedback. The level of this steady state is so low that the amount of thyroid hormone produced is not sufficient to make the normal animal deviate perceptibly from the thyroid-ectomized one.

As the growth curve of the animal suggests, and as has been recognized and emphasized by Gudernatch and other early workers in the field, there is a reciprocal relation between the rate of overall body growth and the rate of metamorphic change. Therefore, we may inquire as to the significance of the endocrine factors in maintaining the high growth rate of the premetamorphic period. It was recognized early that thyroid treatment inhibits growth even as it promotes metamorphosis. Steinmetz (1952 and 1954) studied this phenomenon quantitatively by measuring the growth rate of tadpoles immersed in various concentrations of thyroxine and treated with the goitrogen, propothiouracil (PTU). He showed that distinct inhibition of the growth rate occurs at the lowest concentration of thyroxine measurably affecting leg growth and that the goitrogen stimulates the tadpole's growth. From this he inferred that the level of thyroid hormone in the premetamorphic animal is maintained at an extremely low level, less than that equivalent to one part thyroxine per billion parts of water. Our own unpublished work with thyroxine concentration applied to normal tadpoles fully confirms Steinmetz's results, as illustrated in Fig. 10.3. Thus, it is clear that the low steady state of the PT axis is an essential factor in maintaining the characteristic high growth rate of the premetamorphic period.

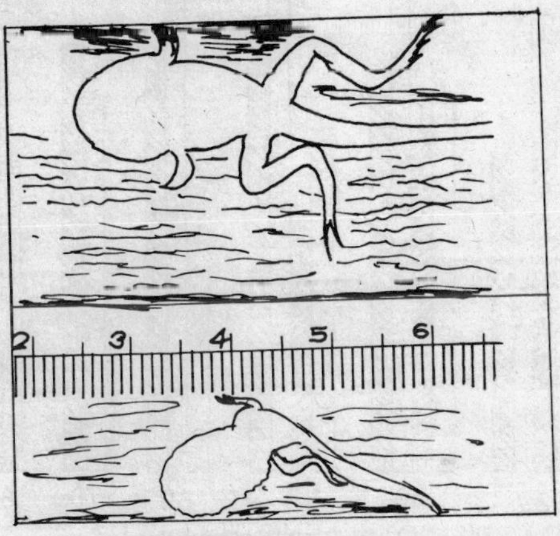

Fig. 10.2. Normal animal (above) compared to experimental animal raised in 3 to 9 ppb thyroxine.

In recent years it has come to be realized that prolactin (or a prolactinlike molecule) plays a significant role in amphibian physiology. Berman *et al.* (1964) and Nicoll *et al.* (1965) reported a stimulation of growth in normal tadpoles by prolactin, and Etkin (1964) and Etkin and Gona (1967) reported such stimulation in hypophy-sectomized animals. These studies indicate an antagonism between thyroid hormone and prolactin. The antithyroid activity of prolactin has been found at low doses to act peripherally and to act as a goitrogen at high dosage level. In any case it appears that the retardation of growth to be seen in the hypophysectomized tadpole is to be ascribed at least in part to the absence of the prolactinlike factor. Further support of this concept of derived from the report of Etkin and Lehrer (1960). They found that the transplanted pituitary reduces growth in hypophysectomized hosts in excess of that of the normal. This suggests that the production of a prolactinlike hormone in the amphibian, like its production in the mammal, is under inhibitory control by the hypothalamus.

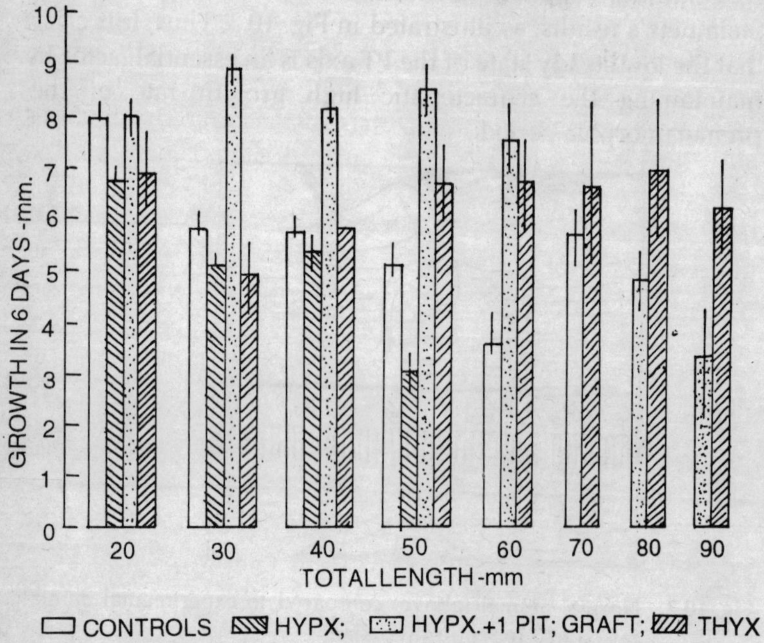

Fig. 10.3. Rate of growth of tadpoles of *Rana pipiens*. All animals of same batch. Individual growth curves kept for each animal and amount of growth in a six-day period as derived from these growth curves is plotted at different stages of development. Controls entered metamorphic climax at 65 to 70 mm. Hypophysectomized animals generally failed to grow above 60 mm.

The concept of the neuroendocrine relationships responsible for the high growth rate and the low metamorphic tendency in the premetamorphic tadpole that emerges from this evidence may be summarized as follows. The hypothalamus is only minimally active in producing a TRF or a prolactin-inhibiting factor. By this activity it only partially retards prolactin production and does not appreciably stimulate TSH production. Consequently, the negative feedback of thyroxine upon the activity of the TSH cells prevails and keeps the PT axis at a very low level. The high level of prolactin and low level of thyroid hormone favour a high growth rate and no appreciable metamorphic change during this period.

The Regulation of the Pattern of Tissue Response

The concept of the low-level steady state of the metamorphic mechanism during the growth phase of the tadpole's life raises most directly the question of how this system is disrupted to permit metamorphosis to take place. Before we can examine this question, however, it is necessary to arrive at some understanding of endocrine activity during metamorphosis so that we can appreciate what the nature of the transition from one condition to the other might be. The clue to this was given by the analysis of the pattern of thyroid activity in relation to the pattern of metamorphic change.

It will be recalled from a preceding section and Figs. 10.1 and 10.2 that the prometamorphic period is characterized by rapid growth of the hind legs and other less conspicuous changes that are spread over about three weeks in *R. pipiens*. This is followed by about a week of climax, during which profound changes in morphology occur. The relation of developmental and histological changes in the thyroid gland to these events was early subjected to extensive studies. The results in different anuran species indicate that during the period of prometamorphosis the relative growth rate of the thyroid is high and the gland reaches a maximum size at the end of this period. The histological picture of moderate increase in cell height accompanied by enlargement of follicles indicates cell activity and storage of hormone. In early climax, however, the epithelium reaches a maximum height and the follicles show a reduction of colloid which, particularly in smaller species, brings on partial collapse of the follicles. This picture was early interpreted as indicating an increasing release of thyroid hormone during prometamorphosis and a very high level of activity at climax (see the rèview of Etkin, 1955; Schrekenberg, 1956). At the end of climax the histological picture shows a return to a flat epithelium and to follicles distended with colloid, thus indicating a deactivation of the thyroid.

Modern studies have attempted to apply recently developed techniques to the analysis of thyroid activity in relation to metamorphosis. These have included chemical studies with and without radioiodine electrical conductivity measurements and enzyme studies. In a general qualitative way these studies have supported the morphological investigations mentioned above since most authors interpret their findings in terms of increasing activity with the approach of metamorphic climax. However, the quantitative refinement that one could wish for from such studies was not forthcoming. For example, Saxen et al. found no clear increase in protein-bound iodine in prometamorphosis and only a twofold increase at climax. Whereas these authors and Kaye reported a large increase in I^{131} uptake, this uptake peaked in mid-or late prometamorphosis rather than at climax. Similarly, the conductivity measurements of Gorbman and Ueda indicated the greatest change in midprometamorphosis and a return to the premetamorphosis level at climax. Flickinger was able to detect thyroxine in peripheral tissues only at climax. Yamamoto followed t_4-deiodinase activity through preisely defined stages of metamorphosis and reported a great increase during prometamorphosis peaking at climax, falling abruptly thereafter. Although the physiological significance of this deiodination is not clear, the parallelism to metamorphic activity is suggestive of changes of hormone level. Unfortunately, the results of Dowling and Razevska are not in agreement with the findings of Yamamoto's study. The reason for this difference remains unexplained. Needless to say, each of these techniques has its complexities of interpretation, the analysis of which would take us far afield at this point. Therefore, without going into detail, the present author may summarize his interpretation that these studies reinforce the concept that prometamorphosis is marked by an increase in thyroid activity, much of which goes into storage of hormone in follicles. However, they fail to clarify in any quantitative way the changes of level of effective hormone. Particularly, they fail to define the effective hormone level responsible for climax changes. It is to be hoped that this admittedly personal evaluation of the results of these methods constitutes a challenge to a critical application of biochemical and biophysical methods to this problem. There is as much to be gained in the clarification of the methods as well as in the understanding of thyroid action in metamorphosis by such a study. A most promising application of modern biochemical methodology to this problem is the attempt of measure T_4 and T_3 levels in the blood directly by chromatographic separation. Preliminary results by this method support the concept of increasing concentration of thyroid hormone during prometamorphosis to a maximal at climax with a decrease thereafter. But a more precise qualification of hormone level is not yet available.

An indirect method for evaluating thyroid function quantitatively is that of studying metamorphic change in young tadpoles, preferably thyroidectomized animals, during immersion in thyroxine solutions of various concentrations. An early study demonstrated that low concentrations (3 to 10 parts thyroxine per billion parts of water) yielded good leg growth, but later metamorphic change was protracted at this hormone level. Higher concentrations (100 to 1,000 ppb) on the other hand yielded climax events at near normal rates but did not permit sufficient time for leg growth to attain normal dimensions before tail resorption was induced. The general inference drawn from these studies was that a normal pattern of metamorphic change could be induced only by an extended period of treatment with a low concentration of T_4 followed by immersion in high concentration. This inference has been repeatedly confirmed in many subsequent studies in this laboratory.

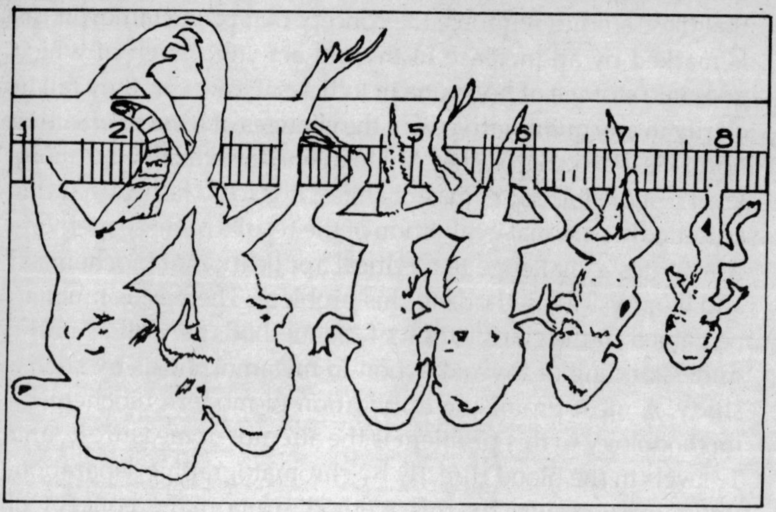

Fig. 10.4. Tadpoles of *R. pipiens*. Second from right is an untreated control fixed three days after emergence of forelegs when tail resorption is half completed. The maximal size attained by this animal was 68 mm. The animal at right was a thyroidectomized animal induced to metamorphose by a patterned application of increasing concentrations of thyroxine after reaching 90 mm. To the left are a series of normal animals similarly treated beginning at various early stages and fixed at midclimax. Note the normal morphology induced by the pattern of increasing concentration of hormone in contrast to the abnormalities induced by single concentrations.

More recently we have attempted to give this concept more precise quantification by studying the relation of level of immersion to rate of a quantifiable event of prometamorphosis (leg growth) and of climax (tail resorption). The graph shows that the maximal rate of leg growth comparable to that in normal animals may be attained in concentrations of 10 to 20 ppb, whereas maximal rates of tail resorption require well over 200 ppb for this induction. The general conclusion that

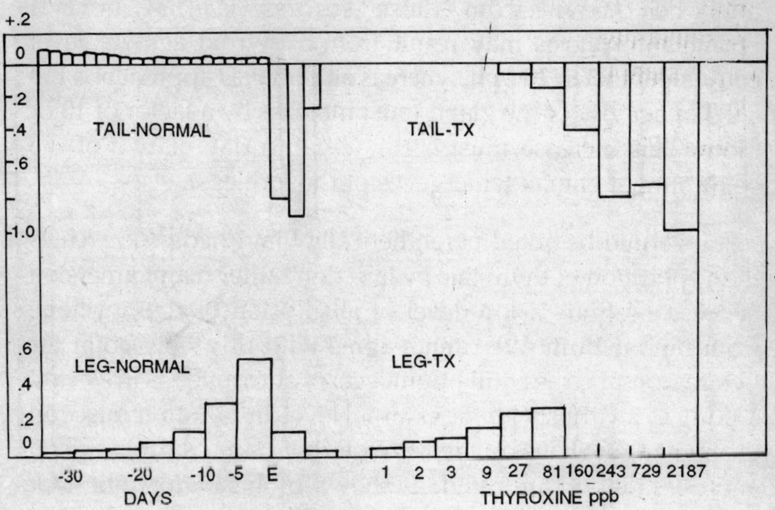

Fig. 10.5. Results of an experiment (right) to test the effect of different concentrations of thyroxine (Tx) on hind-leg growth and tail resorption in half-grown normal tadpoles. For comparison the rates of change of leg and tail in normal metamorphosis are shown at the left. The rates of change are expressed as proportions of body (snout to anus) length to make the data of smaller experimental animals comparable to controls. Each bar for the tail shows the ratio change over a two-day period in normal animals and for days 3 to 5 after exposure to Tx for the experiments. Leg ratios are shown for five-day periods in normals and days 3 to 8 in experimentals. It may be noted that in the normals the legs grow most rapidly in the five-day period before climax begins (E), and tail reduction begins after E and drops by 0.8 of the body length in each of the next two-day periods. In the experimental animals it can be seen that the rate of leg growth produced by 27 ppb thyroxine is about maximum. Animals in concentration higher than 81 ppb do not survive long enough to provide adequate data, although observations shows that their legs do not grow faster than those in animals at 27 ppb. The tail, on the other hand, does not show its maximal rate of reduction until the concentration of 243 ppb is reached. It may be concluded that maximal leg growth such as occurs during late prometamorphosis requires a Tx concentration about one tenth that necessary for the maximal rate of tail resorption.

may be drawn from these studies is that normal prometamorphosis may result from a thyroid activity level equivalent to 3 to 20 ppb, whereas as climax is approached the level of activity of the gland must increase by a factor of 10 or more. This increase must occur within a day or two of the beginning of climax (emergence of the forelegs).

It should be noted parenthetically that Frieden advocates the application of thyroxine by injection rather than immersion because he finds a low level of absorption of the hormone from the medium. We cannot agree with this view-point for two reasons: (1) the equilibrium level of absorption is irrelevant as long as it varies with the external level and (2) after injection the internal level does not remain constant since loss by excretion and degradation are rapid, as shown by his own studies. Of course, immersion studies do not tell us directly what the equilibrium level of hormone is within the animal's body, but Kollros (1963) has shown that the hormone's concentration is maintained at a highly constant level by the immersion method.

This type of evidence then leads to the concept that the level of thyroid hormone action rises progressively during prometamorphosis and undergoes an explosive increase by a factor of at least 10 times at climax. The most immediate question raised by this concept is that of the significance of this changing thyroid hormone level for the induction of the normal sequence and spacing of metamorphic events. We continued here with the endocrine analysis by asking how this pattern of thyroid activity is regulated.

The Control of Level in the Pituitary-Thyroid Anix

Since thyroid activity above the minimal level as found in the hypophysectomized tadpole is dependent upon TSH production, it could be assumed that the activation of the thyroid in metamorphosis is induced by an increase of TSH release from the pituitary. Exploration of the morphology of the pituitary therefore might be expected to yield as clear a picture of activation as did the studies of the thyroid discussed above. There have been repeated studies of this problem, but although a general qualitative agreement with this concept has been reported, the clarification of the quantitative interrelations is not impressive. Recent studies by Kerr (1966) in *Xenopus* and by Kiremidjian and Ortman (1966) in R. *pipiens* have indicated an increase in the supposed TSH cells (identified as Aldehyde fuchsin positive basophils). Yet the amount of the increase, even admitting that the level of activity of a cell is not readily inferred from its staining reaction, does not seem entirely consistent with the inference from the thyroid pattern. Kerr emphasizes that there is no evidence of hormonal discharge in the pituitary at the time of metamorphic climax. The identification of the TSH cell as a basophil is itself uncertain. Several authors have presented evidence pointing to an acidophil as the source of THS. The study of this problem at the electron microscope level by Dent and Gupta (1967) again indicates some increase in the supposed TSH cells but not as great a change at climax as was to be anticipated. The recent finding of the antithyroid activity of prolactin suggests that instead of a simple increase in TSH there may be a shift in proportion of TSH to prolactin production that accounts for the activation of the thyroid gland. Indeed, the criteria for identifying the TSH cell (response to goitrogens, thyroxine, etc.) will have to be reconsidered in the light of a possible role for prolactin in controlling thyroid activity. In any case, it is apparent that further quantitative work will be required before we can specify the nature of the change in the pituitary activity that is responsible for the patterning of thyroid activity during metamorphosis.

But whatever may be the precise nature of the change in the pituitary, it is clear that the signal initiating and controlling the pattern of metamorphosis must come from the pituitary. Therefore, we may ask whether the pituitary is autonomous, i.e., whether it determines its own rate of activity by some sort of built-in' clock mechanism or whether it is subject to some external control. The remarkable fact that throughout the vertebrates, however diverse their brain-pituitary morphology may be, the connection between the two structures is always maintained suggests that the pituitary is under control by the CNS. Furthermore, the clarification of the morphological relationship by the demonstration of the neurosecretory system and the pituitary portal system of blood vessels strongly reinforces this viewpoint. Before the nature of the neurosecretory system was understood, this problem had been attacked in relation to anuran metamorphosis by isolating the gland from the brain. In 1938, Etkin showed that if a tadpole's own pituitary were transplanted in the embryo the animal would usually grow very well but would show at best a retarded and protracted metamorphosis. Nonetheless, the capacity of at least the most successful cases for inducing prometamorphosis was taken to indicate the essential independence of the pituitary from its connection to the brain, a connection then thought of in terms of conventional innervation. In 1940, Uyematsu reported the converse experiment in a toad, removing the region of the embryonic neural plate that gives rise to the infundibular lobes of the hypothalamus and thereby isolating the epithelial pituitary. He also reported his animals to show prometamorphosis, but he found them unable to complete climax changes. This failure he ascribed to a degeneration of the isolated gland at climax. But despite this evidence, which speaks for the concept of an autonomous clock mechanism in the pituitary as the controlling factor in metamorphosis, when the nature of the neurosecretory and the pituitary portal system and hypothalamic influence on TSH production was elucidated in mammals, the need to reexamine this question was recognized. We were able to confirm out earlier work with embryonic grafts and extended it to grafts of differentiated glands placed into hypophysectomized hosts. In these studies the animals were maintained into climax, and it was found that the hosts to successful grafts, though often completing prometamorphic development, invariably entered a metamorphic stasis at climax. Clearly, then, graft activity was capable of activating the thyroid to the low levels responsible for prometamorphic change but not to the high level necessary for climax events. It was apparent that the failure at climax was not due to a degeneration of the gland at that time since the gland was found to be maintained morphologically and also was seen to continue the production of its growth and pigmentation-regulating factors in spite of the metamorphic stasis. We were also able to confirm Uyematsu's findings of similar capacities in pituitaries isolated by removing part of the embryonic hypothalamic region.

There have been many reports in the last 10 years confirming the necessity of the presence of the hypothalamus for the completion of metamorphosis. However, Guardabassi (1961) reported completion of metamorphosis after removal of the hypothalamus. It should be understood that failure of metamorphosis in short-term experiments or in experiments in which the animals fail to grow vigorously does not necessarily imply hypothalamic control of TSH function since injury or starvation by themselves can inhibit metamorphosis. However, in our experiments and those of Remy and Bounhiol there can be no doubt of the vigour of the animals since many achieved gigantic size without completing metamorphosis. At least in these experiments, then, it is clear that isolation of the pituitary from the hypothalamus specifically inhibits it from attaining a high level of TSH activity. It is possible that the discordant results reported by Guardabassi are the consequence of incomplete removals accompanied by regenerative repair, which is possible in the tadpole's nervous system. In the opinion of this author it appears to be clearly established that the high level of thyroid activity necessary for metamorphic climax can be maintained by a pituitary gland only if it has its normal connection with the brain. In urodeles also it is clear that the metamorphic process that corresponds to climax in anurans depends upon normal hypothalamic relations with the pituitary. A barrier inserted between these two structures (see frontispiece D) effectively prevents metamorphosis, provided it completely separates them. A low prometamorphic level, however, can be maintained by the isolated pituitary in anurans.

On the other hand, some evidence indicates a considerable measure of independence of the pituitary TSH production from hypothalamic control in amphibians. Dent (1966) reported that molting in the newt, in which this process is known to be dependent upon the thyroid, is maintained by pituitaries isolated from the brain. Iodine metabolism, though reduced in-toads with pituitary grafts, is higher than in hypophysectomized animals. We regard such evidence as consistent with that derived from metamorphic studies where the isolated pituitary is found to sustain prometamorphic change. It is possible that the role of thyroid in the molting of these amphibians is merely permissive. Though necessary at a minimal level, it does not determine the rhythm of molting by its variation.

One is tempted to infer from the evidence given above that the pituitary TSH system is autonomous for low levels of activity and requires hypothalamic cooperation only for the highly activated state. However, if we think of the hypothalamic-pituitary interrelationship in terms of neurosecretion rather than conventional innervation it is apparent that inference is not justified. Presumably, the significance of the pituitary portal venous system lies in the greater efficiency it provides for the transfer of chemicals produced in hypothalamic neurons to the cells of the anterior pituitary. Thus, it appears likely that though this system of direct transfer of neurosecretion may be essential for maximal stimulation of the pituitary, a lower level of TSH activity may be maintained by the transfer of neurosecretion through the systemic circulation. Such a low level of stimulation could reach a graft wherever, it was located. The evidence on brain removal mentioned above either does not include the areas presumed to contain the relevant pericarya or is too fragmentary and incompletely analyzed (Voitkevich and Guardabassi) to permit judgment. Further studies should clarify this critical point. At present it appears most logical to accept the possible transfer of hypothalamic substances in diluted form through the systemic circulation as the explanation of the low and moderate levels of TSH function reported in isolated pituitaries.

Of course, a most direct approach to this question could be derived from experiments with extracts of the hypothalamus that have thyrotropin-releasing properties (TRF). Such TRF preparations have been developed for mammals in several laboratories. However, Bowers and Schally, working with their TRF preparation, failed to find evidence of TSH release in the tadpole (personal communication). In our laboratory, work with a TRF prepared by Dr. Roger Guillemin that was potent in tests on rodents likewise failed to elicit activation of metamorphosis, even when large amounts were injected directly into the heart. Presumably the chemistry of the amphibian TRF must differ from that in mammals, but no adequate studies of an amphibian TRF are yet available. The possibility should also kept in mind that a shift in the prolactin-TSH balance rather than a simple release of TSH may be necessary for metamorphic activity.

Our consideration of the hypothalamic-pituitary-thyroid-tissue axis in relation to metamorphosis thus leads us to the concept that at the beginning of prometamorphosis the hypothalamus initiates the changes, in some way stimulating the pituitary-thyroid axis. This activation process builds up during prometamorphosis to a subtotal activation of the axis by the beginning of climax. At the end of metamorphosis the hypothalamus apparently stops its stimulation of the PT axis since the latter becomes inactive, although we have no direct evidence on this point. Such a concept raises the question of the control of the hypothalamus: Does it, rather than the pituitary, contain the "clock" mechanism that determines the timing and pattern of metamorphosis?

The Activation of the Metamorphic Mechanism

Sine the evidence indicates that the control of the metamorphic process depends upon the hypothalamic neurosecretory apparatus, it was desirable to investigate the development of this mechanism. Wilson et al. (1957) reported on the morphology of this system in the western tree frog (*Hyla regilla*) and showed that the classical neurosecretory material of the posterior lobe (see above) appears early in development. However, they did not follow the differentiation of the portal vessels and medium eminence in detail. When this was studied in *R. pipiens.* in correlation with metamorphic change, it was found that in the premetamorphic tadpole these parts of the neurosecretory apparatus are poorly differentiated. The anterior lobe of the pituitary is broadly adherent to the thin floor of the hypothalamic lobes with a diffuse network of capillaries lying between the two. The capillaries of this network derive their blood from hypothalamic arteries and feed into the capillary system of the anterior pituitary. During prometamorphosis this system differentiates in the neural floor, becoming shorter and thicker as the capillaries sink into the neural substance. The anterior lobe separates from the neural tissue except at the anterior tip, where it remains attached by the few large venous channels that drain the primary capillary bed. These constitute the pituitary portal veins draining the now-differentiated medium eminence as found in the adult; when the animal enters climax, this medium eminence system is well advanced although not completely differentiated.

It is evident that this morphological development is entirely consistent with the concept that some hypothalamic substance is fed from the neurosecretory fibers in the hypothalamic stalk into the capillary bed of the anterior lobe. In the premetamorphic animal this system, being only diffusely organized, presumably carries a minimum of hypothalamic substance. As the system matures during prometamorphosis and the medium eminence differentiates, the amount of material transported increases to reach a level sufficient to achieve the high-level activation of the PT axis at metamorphic climax.

If this hypothesis is valid, then the question of hypothalamic control could be studied by examining the differentiation of the median eminence region in the thyroidectomized larva. If the hypothalamus is autonomous in its differentiated, the median eminence should mature in spite of the failure of the thyroidless larva to metamorphose. The examination of this question

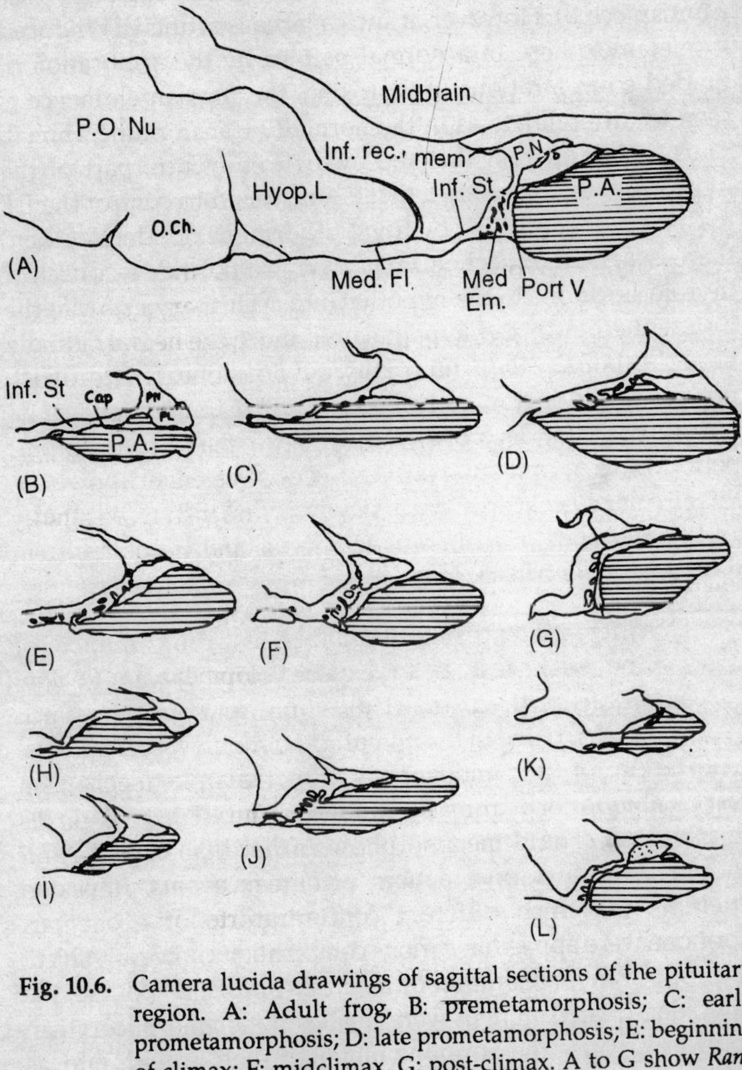

Fig. 10.6. Camera lucida drawings of sagittal sections of the pituitary region. A: Adult frog, B: premetamorphosis; C: early prometamorphosis; D: late prometamorphosis; E: beginning of climax; F: midclimax. G: post-climax. A to G show *Rana pipiens*. H: Early prometamorphosis; I: late prometamorphosis; J: one day before E. H to J show *Rana sylvatica*. K: mid-prometamorphosis; L: early climax. K to L show *Bufo americanum*.

showed that this was not so. In the thyroidectomized larva the median eminence region retains its larval character (see frontispiece B). However, if such a larva is artificially induced to metamorphose in a normal pattern by the application of graded concentrations of thyroxine, the median eminence is seen to differentiate as in the normally transforming animal. Thus, it is apparent that the median eminence, part of the hypothalamic mechanism that is presumed to control the PT axis, is itself controlled by thyroid hormone. Evidently, then, there is a circular mechanism involving a positive feedback of thyroid hormone to the hypothalamus. This suggests that the differentiation of the hypothalamic mechanism is gradually brought about by the action of thyroid hormone. This feedback process starts in the early larva at an extremely slow rate because of the low level of thyroid activity in the premetamorphic period. Since it contains a positive feedback element, however, this circular system accelerates slowly at first then more rapidly and finally ends in a burst of activity as any circular system with positive feedback must.

This intriguing possibility could be tested in a number of ways. One such test is to expose early larvae to low concentrations of thyroxine (1 to 9 ppb) for sometime. Such exogenous thyroxine in this hypothesis, should, accelerate the maturation of the endogenous hypothalamic mechanism responsible for activating the animal's own PT axis. Animals thus treated should metamorphose earlier than controls and should show a normal pattern of climax events. Repeated attempts in which different concentrations of exogenous hormone were applied in various combinations however, failed, to produce an acceleration of metamorphosis. Of course, the thyroxine-treated animals were smaller at metamorphosis than controls because the hormone inhibited their growth rate, as explained above. This result raised the possibility that the hypothalamic system, as certain other neural elements, does not acquire its sensitivity to thyroxine until late in the premetamorphic period. An experimental analysis of this point showed that it was indeed so. Animals (*R. pipiens*) treated with graded concentrations of thyroxine from early larval stages and thus brought into mid-climax much earlier than normal failed to show any metamorphological advance of the median eminence region over the larval condition. On the other hand, animals that were so treated as to be only slightly precocious did show such development. Thus, it is clear that until a short time before the normal animals enter prometamorphosis, the median eminence region, and presumably the rest of the hypothalamic neurosecretory apparatus, is sensitive to positive thyroid feedback. The feedback cycle thus begins about a week before a commencement of metamorphosis in normal development and builds up quickly through prometamorphosis to climax.

Thus, we are led to consider that the timing of the beginning of metamorphosis is regulated by a clock mechanism in the hypothalamus that times the acquisition of hormone sensitivity in that organ. On the other hand, the pattern of thyroid hormone build up that, as we have seen previously, determines the pattern of metamorphic change is determined by the positive feedback system of thyroxine to the hypothalamus. In order to form a more comprehensive picture of the mechanism regulating development in larval life of the frog we must now add the newly acquired knowledge of the role of prolactin in promoting body growth and inhibiting metamorphosis in tadpoles as discussed above. Since prolactin is under inhibitory control by the hypothalamus, the positive feedback of T_4 to the neurosecretory system presumably inhibits prolactin activity while at the same time stimulating the PT axis. In Fig. 10.9 an attempt is made to depict the interaction of these elements in the life history of the anuran larva.

Metamorphosis as a Development Pheomenon

I propose in this section to speculate more broadly than, in the stress of scientific investigation and writing, scientists commonly permit themselves to do. Since the ideas expressed are very general, no detailed documentation will be offered.

Evolution and Developmental Mechanisms in Amphibians

Although a few fossil larvae are known, they contribute little to an understanding of the evolution of metamorphosis beyond the fact that it was present is paleozoid labyrinthodonts. General biological considerations, however, permit some reasonable speculations. Metamorphosis in urodeles is a developmental adaptation permitting the change from water to land environment. The basic mode of feeding by predation and the associated locomotor mechanisms, however, remain unchanged. Presumably this was the situation is primitive amphibians. In anurans, on the other hand, the larva generally feeds as a free-swimming scavenger, herbivore, or filter feeder. In metamorphosis it changes to a predator which depends upon powerful leg action for its saltatory locomotion. The significance of the prometamorphic phase in anuran transformation appears to lie chiefly in providing time for the legs, whose development had been repressed during the larval phase, to grow before the transition to land is made at metamorphic climax. It urodeles the limbs develop in the way usual among vertebrates, during embryonic stages, and are not involve in metamorphosis.

This evolutionary process must, of course, have operated through modification of the mechanics of development. From this point of view, the changes of prometamorphosis, principally limb growth, have been brought into the metamorphic picture by the evolution of a developmental mechanism separating limb differentiation from the general embryonic processes and suppressing it during the premetamorphic period. The action of thyroid hormone, then, appears as the factor which removes this initial inhibition of limb development. We see here an example of a phenomenon appearing frequently in the mechanics of development. This is that control of development is effected by a balance between inhibiting and disinhibiting factors operating on a protoplasmic mechanism with very broad developmental capacities. We see this type of development control in gene action at the molecular level, in fertilization phenomena, and in field effects in organ determination.

At the level of endocrine action we again see that developmental control is effected by a balance between inhibition and disinhibition rather than by simple stimulation. The PT axis is kept at a low level of activity in the growth phase of the tadpole's development by negative feedback, and at metamorphosis its activation is brought about by disinhibition of this feedback by hypothalamic action. It is possible that the hypothalamic controlling mechanism of the pituitary has been inhibited for the duration of the larval period by some evolutionary change in the genome. In that case what we have described above as the positive feedback of thyroxine to the hypothalamus would be another instance of the disinhibition of development by thyroxine.

The complexity of the push-pull type of interaction in governing metamorphosis is further emphasized by the discovery of the role of prolactin as a thyroid antagonist in amphibian development. Whether this substance acts at the peripheral level or as a goitrogen or in both ways its role again emphasizes that development is controlled by a dynamic balance of plus and minus factors. A fascinating aspect of this particular interaction is the manner in which it appears to have been seized upon as the mechanism for the evolution of the "second metamorphosis" in the common newt. After a first metamorphosis that produces the land form, this animal returns to the water for breeding in a second period of transformation (see frontispiece A). This second metamorphosis has been shown to be under the influence of prolactin. We must now view this change as a shift in the balance between the pituitary factors, prolactin and TSH, rather than as an activation of one of them. First metamorphosis is induced by a shift in favour of TSH, second metamorphosis by a return to the predominance of prolactin. What brings on this second shift is yet to be explored.

Insect and Amphibian Metamorphosis

The analogy between insect metamorphosis and that of amphibians has often been commented upon. In both, the brain acting through a neuro-secretary mechanism controls the endocrine apparatus which governs tissue response through hormone release. Until recently one feature of the insect system seemed to be lacking in vertebrates. In insects the hormone of the corpus allatum acts as a modulator inhibiting the action of ecdysone. We now see, however, that there is an analogous situation in amphibians, with prolactin playing the role of modular to thyroid activity. Perhaps the analogy will suggest ways in which factors that are found to regulate balance in one system can be applied to the other.

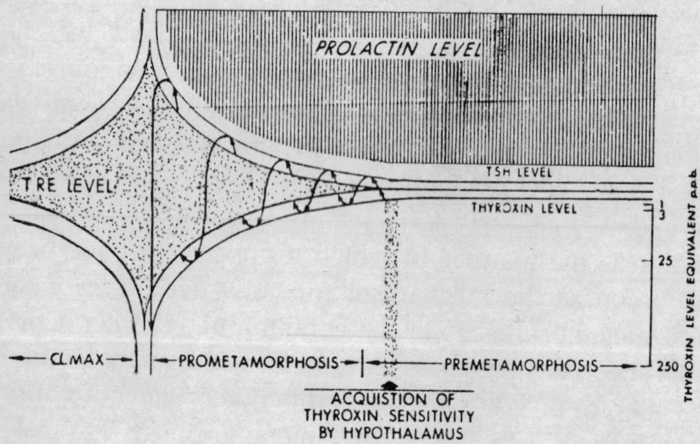

Fig. 10.7. Diagram illustrating the interaction of endocrine factors in determining the time and pattern of anuran metamorphosis.

Metamorphosis and Puberty

A final area of speculation deserving comment is the analogy between amphibian metamorphosis and puberty in mammals. There is striking similarities in the mechanism of sexual regulation and metamorphosis. In each, the hypothalamic neurosecretory system appears dominant, controlling the pituitary (gonadotropin or TSH) which controls the basic endocrine gland (gonal or thyroid). This is turn regulates the response of the peripheral tissues. Within this system, likewise, there is a negative feedback relation which makes this, too, a balanced system.

In the prepuberal mammal in the amphibian larva the lower levels of the control sequence are capable of response much before they are actually called upon to act. Thus, the sex accessories and secondary sex characters can respond to gonadal stimulation long before puberty in most mammals. Likewise the gonads and the pituitary show early responsiveness. The critical factor determining the time of puberty then lies in the activation of the hypothalamus just as it does in metamorphosis. Beyond this point, however, the analogy appears to break down. Property placed lesions in the hypothalamus have been shown to accelerate puberty so that the hypothalamic gonadotropic releasing mechanism appears to be under inhibitory control of other neural centers. This has never been shown to apply to amphibian metamorphosis. On the other hand no positive feedback system of hormones to the hypothalamus has been desçribed for the mammal as it has for the amphibian larva. Could this be because the possibilities suggested by the analogy have never been explored?